C0-ACH-360

The New York Times

IN THE HEADLINES

Adapting to Climate Change

THE NEW YORK TIMES EDITORIAL STAFF

Published in 2020 by New York Times Educational Publishing
in association with The Rosen Publishing Group, Inc.
29 East 21st Street, New York, NY 10010

First Edition

Cataloging-in-Publication Data
Names: New York Times Company.
Title: Adapting to climate change / edited by the New York Times editorial staff.
Description: New York : The New York Times Educational Publishing, 2020. | Series: In the headlines | Includes glossary and index.
Identifiers: ISBN 9781642823004 (library bound) | ISBN 9781642822991 (pbk.) | ISBN 9781642823011 (ebook)
Subjects: LCSH: Climatic changes—Juvenile literature. | Climatic changes—Effect of human beings on—Juvenile literature. | Climatic changes—Government policy—Juvenile literature. | Global warming—Juvenile literature. | Nature—Effect of human beings on—Juvenile literature.
Classification: LCC QC903.15 A337 2020 | DDC 363.738'74—dc23

Manufactured in the United States of America

On the cover: A resident wades through water at high tide in the village of Abarao on South Tarawa, one of the atolls of Kiribati, March 26, 2016; Josh Haner/The New York Times.

Contents

CHAPTER 4

Climate Skepticism Becomes a Partisan Opportunity

CHAPTER 5

Technical Solutions to Climate Crisis

CHAPTER 6

Action and Inaction in the Political Sphere

Introduction

CLIMATE CHANGE IS old news, confirmed over a century of scientific study. Its effects are the future. As the Intergovernmental Panel on Climate Change noted in its Fifth Assessment Report, released in 2018, we are now on an irreversible path to 1.5 degrees Celsius warming above pre-industrial levels, and we see the results all around us.

Climate science is based on old hypotheses, elaborated on and confirmed over time. In 1896, Svante Arrhenius concluded (based on John Tyndall's 1850s research on the greenhouse effect) that the carbon dioxide emitted by coal-burning industry could one day increase global temperatures. By the 1950s, atmospheric scientist Charles Keeling set out to measure those emissions and set the benchmark for measuring the upward trend of atmospheric CO2 — known as the Keeling Curve.

That discovery spurred climatologists to devise computer models predicting the effects of that carbon dioxide on the global climate. The results were concerning: between three and nine degrees Fahrenheit of average warming, higher at the poles, with resulting ice cap melting and increased severity of droughts and storms. NASA scientist James E. Hansen brought that research and those warnings to Congress in 1988. His gloomy forecasts sparked thirty years of debate. During that time, warming continued, and prediction became fact: the ten hottest years ever recorded took place in the last two decades.

The weather effects of climate change are unequally distributed. Some areas get torrential rains and others get searing drought. Extreme winter weather in some regions occurs in the same year as prolonged heat waves in others. The damage, sadly, is also unequally distributed. Poorer regions are seeing their way of life destroyed, making wars more desperate and accelerating the refugee crisis.

JOSH HANER/THE NEW YORK TIMES

The dry bed of Lake Poopó in Bolivia, where steady warming over the last 30 years evaporated what little water was left.

To prevent warming above 1.5 degrees Celsius, the I.P.C.C. calls for international action to reduce emissions of greenhouse gases. Unfortunately, those reductions require comprehensive changes to global energy, transportation and food infrastructure. Efforts also require shifts in energy demand and efforts to remove carbon dioxide from the atmosphere. The status quo means warming closer to 2 degrees Celsius, which could expose hundreds of millions more people to instability, famine and disease.

Current trends are against that goal. Rich nations, which are responsible for the majority of global emissions, are falling back on their prior commitments due to logistical challenges or changing political trends. Emerging economies are increasingly dependent on coal, shaped by China's influence over global development. Though the United States and China — the two largest contributors to climate change — have bases of support for action, this period of nationalistic economic competition risks a deadly delay of that action.

Fortunately, many argue that the technical requirements for limiting climate change are more feasible than ever. Renewable wind and solar power has improved, solutions have been found for logistical problems and many smaller innovations show promise. Solutions are in sight, and the United Nations argues that many of them can even play a role in fighting global poverty. Fighting climate change is possible. The problem is a lack of political will; those with decision-making power lack the awareness, interest or incentive to make a difference.

Some people argue against the consensus that human activity is the leading factor in climate change. Why do people doubt? On the one hand, the process of scientific inquiry requires it and welcomes dissenting views. N. Ángel Pinillos's article, included in this collection, offers helpful insights on how to use our doubts effectively, to learn and act. On the other hand, many doubt because it is in their political or economic self-interest to do so, such as fossil fuel lobbies and the politicians they support. Rich nations with fossil fuel industries to protect can afford to deny the science. Poor nations do not have that luxury.

Given our government's haphazard action on climate change, many have looked to individual actions and market solutions to fill the gap. However, research suggests that those components alone cannot meet the scale required to reduce emissions. Public investment has proven essential to organizing available tools and spurring the invention of new ones. Looking ahead, policymakers in the United States are following examples in Costa Rica and New Zealand with an ambitious proposal: a Green New Deal, which would invest trillions of dollars in a complete carbon-neutral infrastructure.

The future of climate change is a political choice for all of us to make. The discipline of ecology often teaches us to focus on connections, even between distant things. Climate change highlights such connections between our lifestyles, our environment, our knowledge and our ways of acting together. In that spirit, Dr. James Hansen suggests that a necessary step for fixing our climate is restoring democracy itself.

CHAPTER 1

Climate Science Evolves, Refines Its Message

In 1988, Dr. James E. Hansen publicly addressed Congress to describe climate change's causes and possible effects. Those statements were based on climate science established as early as the 19th century research on the greenhouse effect, caused by carbon dioxide and other gases. This work was supported by Charles Keeling's later research on rising atmospheric carbon dioxide. The mathematical models and measurement tools have evolved, but the core predictions have come to pass: warming occurred within Dr. Hansen's expected range, and was most pronounced in the polar regions.

Global Warming Has Begun, Expert Tells Senate

BY PHILIP SHABECOFF | JUNE 24, 1988

WASHINGTON, JUNE 23 — The earth has been warmer in the first five months of this year than in any comparable period since measurements began 130 years ago, and the higher temperatures can now be attributed to a long-expected global warming trend linked to pollution, a space agency scientist reported today.

Until now, scientists have been cautious about attributing rising global temperatures of recent years to the predicted global warming caused by pollutants in the atmosphere, known as the "greenhouse effect." But today Dr. James E. Hansen of the National Aeronautics and Space Administration told a Congressional committee that it was 99 percent certain that the warming trend was not a natural variation but was caused by a buildup of carbon dioxide and other artificial gases in the atmosphere.

AN IMPACT LASTING CENTURIES

Dr. Hansen, a leading expert on climate change, said in an interview that there was no "magic number" that showed when the greenhouse effect was actually starting to cause changes in climate and weather. But he added, "It is time to stop waffling so much and say that the evidence is pretty strong that the greenhouse effect is here."

If Dr. Hansen and other scientists are correct, then humans, by burning of fossil fuels and other activities, have altered the global climate in a manner that will affect life on earth for centuries to come.

Dr. Hansen, director of NASA's Institute for Space Studies in Manhattan, testified before the Senate Energy and Natural Resources Committee.

SOME DISPUTE LINK

He and other scientists testifying before the Senate panel today said that projections of the climate change that is now apparently occurring mean that the Southeastern and Midwestern sections of the United States will be subject to frequent episodes of very high temperatures and drought in the next decade and beyond. But they cautioned that it was not possible to attribute a specific heat wave to the greenhouse effect, given the still limited state of knowledge on the subject.

Some scientists still argue that warmer temperatures in recent years may be a result of natural fluctuations rather than human-induced changes.

Several Senators on the Committee joined witnesses in calling for action now on a broad national and international program to slow the pace of global warming.

TRAPPING OF SOLAR RADIATION

Senator Timothy E. Wirth, the Colorado Democrat who presided at the hearing today, said: "As I read it, the scientific evidence is compelling: the global climate is changing as the earth's atmosphere gets warmer. Now, the Congress must begin to consider how we are going to slow or halt that warming trend and how we are going to cope with the changes that may already be inevitable."

Mathematical models have predicted for some years now that a buildup of carbon dioxide from the burning of fossil fuels such as coal and oil and other gases emitted by human activities into the atmosphere would cause the earth's surface to warm by trapping infrared radiation from the sun, turning the entire earth into a kind of greenhouse.

DON HOGAN CHARLES/THE NEW YORK TIMES

Dr. James E. Hansen at the Goddard Institute for Space Studies.

If the current pace of the buildup of these gases continues, the effect is likely to be a warming of 3 to 9 degrees Fahrenheit from the year 2025 to 2050, according to these projections. This rise in temperature is not expected to be uniform around the globe but to be greater in the higher latitudes, reaching as much as 20 degrees, and lower at the Equator.

The rise in global temperature is predicted to cause a thermal expansion of the oceans and to melt glaciers and polar ice, thus causing sea levels to rise by one to four feet by the middle of the next century. Scientists have already detected a slight rise in sea levels. At the same time, heat would cause inland waters to evaporate more rapidly, thus lowering the level of bodies of water such as the Great Lakes.

Dr. Hansen, who records temperatures from readings at monitoring stations around the world, had previously reported that four of the hottest years on record occurred in the 1980's. Compared with a 30-year base period from 1950 to 1980, when the global temperature averaged 59 degrees Fahrenheit, the temperature was one-third of a degree higher last year. In the entire century before 1880, global temperature had risen by half a degree, rising in the late 1800's and early 20th century, then roughly stabilizing for unknown reasons for several decades in the middle of the century.

WARMEST YEAR EXPECTED

In the first five months of this year, the temperature averaged about four-tenths of a degree above the base period, Dr. Hansen reported today. "The first five months of 1988 are so warm globally that we conclude that 1988 will be the warmest year on record unless there is a remarkable, improbable cooling in the remainder of the year," he told the Senate committee.

He also said that current climate patterns were consistent with the projections of the greenhouse effect in several respects in addition to the rise in temperature. For example, he said, the rise in temperature is greater in high latitudes than in low, is greater over continents than oceans, and there is cooling in the upper atmosphere as the lower atmosphere warms up.

"Global warming has reached a level such that we can ascribe with a high degree of confidence a cause and effect relationship between the greenhouse effect and observed warming," Dr. Hansen said at the hearing today, adding, "It is already happening now."

Dr. Syukuro Manabe of the Geophysical Fluid Dynamics Laboratory of the National Oceanic and Atmospheric Administration testified today that a number of factors, including an earlier snowmelt each year because of higher temperatures and a rain belt that moves farther north in the summer means that "it is likely that severe mid-continental summer dryness will occur more frequently with increasing atmsopheric temperature."

A TASTE OF THE FUTURE

While natural climate variability is the most likely chief cause of the current drought, Dr. Manabe said, the global warming trend is probably "aggravating the current dry condition." He added that the current drought was a foretaste of what the country would be facing in the years ahead.

Dr. George Woodwell, director of the Woods Hole Research Center in Woods Hole, Mass., said that while a slow warming trend would give human society time to respond, the rate of warming is uncertain. One factor that could speed up global warming is the wide-scale destruction of forests that are unable to adjust rapidly enough to rising temperatures. The dying forests would release the carbon dioxide they store in their organic matter, and thus greatly speed up the greenhouse effect.

SHARP CUT IN FUEL USE URGED

Dr. Woodwell, and other members of the panel, said that planning must begin now for a sharp reduction in the burning of coal, oil and other fossil fuels that release carbon dioxide. Because trees absorb and store carbon dioxide, he also proposed an end to the current rapid clearing of forests in many parts of the world and "a vigorous program of reforestation."

Some experts also believe that concern over global warming caused by the burning of fossil fuels warrants a renewed effort to develop safe nuclear power. Others stress the need for more efficient use of energy through conservation and other measures to curb fuel-burning.

Dr. Michael Oppenheimer, an atmospheric physicist with the Environmental Defense Fund, a national environmental group, said a number of steps can be taken immediately around the world, including the ratification and then strengthening of the treaty to reduce use of chlorofluorocarbons, which are widely used industrial chemicals that are said to contribute to the greenhouse effect. These chemicals have also been found to destroy ozone in the upper atmosphere that protects the earth's surface from harmful ultraviolet radiation from the sun.

Who Cares About a Few Degrees?

BY ANDREW C. REVKIN | DEC. 1, 1997

SCIENTISTS AND OTHER EXPERTS who have spent years trying to get people concerned about the prospect of global warming have always faced a central problem: temperatures change all the time.

From summer to winter, Minneapolis goes from 90-degree heat waves to 10-below deep freezes. At the end of a spell of Indian summer, Manhattan can go from 65 degrees to 25 in a few hours.

So why should anyone get excited about a global rise of a few degrees in a few decades?

But the debate before negotiators in Kyoto, Japan, is about climate, not weather. It is about long-term shifts in patterns of weather, not seasonal or day-to-day shifts from rain to sun, cold to heat.

Understanding the difference between weather and climate is the first step in understanding why many scientists are predicting that changes that seem trivial in terms of any single day — a change, say, of five degrees Fahrenheit — could have large impacts on many facets of life when those changes are on a global scale.

Global warming would not change the range of weather experienced day to day, but it would increase the odds of having weather that is considered troublesome by 20th-century standards: summer droughts, winter deluges, hurricanes and the like.

At the heart of the theory that has dozens of nations poised to act to lessen the threat is the long-established idea that earth's atmosphere behaves like the roof of a greenhouse.

The atmosphere was first compared to a "glass vessel" in 1827 by the French mathematician Jean-Baptiste Joseph Fourier. He recognized that the air circulating around the planet lets in sunlight — as a greenhouse's glass roof does — but prevents some of the resulting warmth from leaving.

If the air had no heat-trapping effect, the heat from the sun would quickly radiate back into space, leaving the planet with a surface temperature of nearly zero degrees Fahrenheit.

In the 1850's, a British physicist, John Tyndall, took things further and tried to measure the heat-trapping properties of various components of the atmosphere. Surprisingly, the two most abundant gases — nitrogen and oxygen — turned out to have no heat-trapping ability. Ninety-nine percent of the atmosphere has no insulating properties at all. It is all up to a few trace gases — mainly water vapor and carbon dioxide — to keep the planet cozy. If the air did not contain carbon dioxide, the planet would be some 20 degrees cooler. Without water vapor, it would be a deep-frozen ball of ice.

By the 1890's, scientists had figured out that the great blossoming of combustion in the Industrial Revolution had the potential to change the atmosphere's load of carbon dioxide. The idea was summarized succinctly in the April 1896 issue of The London, Edinburgh and Dublin Philosophical Magazine by a Swedish chemist, Svante Arrhenius, who wrote, "We are evaporating our coal mines into the air."

Precise monitoring of carbon dioxide concentrations since the 1950's has shown a relentless upward trend. Some of the carbon dioxide has gone into the ocean, and some has been absorbed by growing trees, but the amount in the air has continued to rise.

There has been a simultaneous rise in the planet's average temperature, although at a far slower — and more uneven — pace. Other factors appear to have acted like a buffer, including a rise in the amount of sooty particles in the air; they act something like a parasol, reflecting some of the sun's energy back into space before it can warm things up.

Nonetheless, many scientists, using computer models, say that they can account for buffering mechanisms and still see problems ahead, particularly if significant cuts are not made in the rates at which petroleum and coal are burned.

Indeed, many of their calculations indicate that the necessary cuts in emissions of carbon dioxide would actually have to be far greater than even the most ambitious targets on the negotiating table in Kyoto.

There has also been a rise in the atmosphere's burden of methane and nitrous oxide from sources related to human activities, and these gases also trap heat.

Even though the greenhouse gases exist in only trace amounts — they are measured in parts per million and, in some cases, parts per trillion — they exert a powerful influence on the temperature of the planet. So a tiny change in their concentrations can cause a big change in the way the atmosphere behaves.

Dr. John W. Firor, an atmospheric scientist at the National Center for Atmospheric Research, likes to compare the situation to that of a corporation that is vulnerable to a takeover. A change of a couple of shareholders' votes can mean the difference between survival and getting swallowed up. It is a "highly leveraged situation," he said, using the parlance of Wall Street.

He and many other scientists say the risks of meddling significantly with the insulating atmospheric greenhouse are simply too great to continue on the current course, adding more than seven billion tons of carbon dioxide and other heat-trapping gases to the air each year.

But they acknowledge that many of the feedback loops and connections between the components of Earth's atmosphere, oceans, ice caps and, ultimately, climate are complex and remain poorly understood. The computer models with which scientists are projecting the range of future consequences from this rise are still a relatively crude representation of the real world of clouds, ocean currents, jet streams and other complexities.

Some consistent critics of the projections say the models are rife with weaknesses — particularly in predicting how warming could affect cloud cover and in how solar energy moves from the surface to the highest levels of the atmosphere and then out into space.

An increase in cloudiness could act like panels on a greenhouse roof, countering the heat-trapping effect of the greenhouse gases by reflecting the sun and leading to cooling.

Dr. Richard S. Lindzen, a professor of meteorology at the Massachusetts Institute of Technology and a consistent skeptic on the perils of climate change, said the negotiations in Kyoto were mostly focused on bolstering the resumes of diplomats.

Even if the models are correct — which Dr. Lindzen doubts — the cuts in greenhouse gases on the table are so small that they will hardly matter, he said.

"Climate always changes, whether man does anything about it or not," Dr. Lindzen said. "Now any changes will be attributed to policy, not nature."

But many of his colleagues disagree, saying that the models have held true despite continuing efforts to weed out spurious results.

And many of them say it is essential to act now. Dr. Stephen H. Schneider, a Stanford University biology professor, said that in the 20th century, there had already been a distinct warming trend and a change in precipitation patterns that was hard to ascribe to anything other than the manmade increase in greenhouse gases.

"Is this nature being perverse or is it us?" Dr. Schneider said. "The only way to prove it for sure is hang around 10, 20 or 30 more years, when the evidence would be overwhelming. But in the meantime, we're conducting a global experiment. And we're all in the test tube."

A Scientist, His Work and a Climate Reckoning

TEMPERATURE RISING | BY JUSTIN GILLIS | DEC. 21, 2010

MAUNA LOA OBSERVATORY, HAWAII — Two gray machines sit inside a pair of utilitarian buildings here, sniffing the fresh breezes that blow across thousands of miles of ocean.

They make no noise. But once an hour, they spit out a number, and for decades, it has been rising relentlessly.

The first machine of this type was installed on Mauna Loa in the 1950s at the behest of Charles David Keeling, a scientist from San Diego. His resulting discovery, of the increasing level of carbon dioxide in the atmosphere, transformed the scientific understanding of humanity's relationship with the earth. A graph of his findings is inscribed on a wall in Washington as one of the great achievements of modern science.

Yet, five years after Dr. Keeling's death, his discovery is a focus not of celebration but of conflict. It has become the touchstone of a worldwide political debate over global warming.

When Dr. Keeling, as a young researcher, became the first person in the world to develop an accurate technique for measuring carbon dioxide in the air, the amount he discovered was 310 parts per million. That means every million pints of air, for example, contained 310 pints of carbon dioxide.

By 2005, the year he died, the number had risen to 380 parts per million. Sometime in the next few years it is expected to pass 400. Without stronger action to limit emissions, the number could pass 560 before the end of the century, double what it was before the Industrial Revolution.

The greatest question in climate science is: What will that do to the temperature of the earth?

Scientists have long known that carbon dioxide traps heat at the surface of the planet. They cite growing evidence that the inexorable rise of the gas is altering the climate in ways that threaten human welfare.

Fossil fuel emissions, they say, are like a runaway train, hurtling the world's citizens toward a stone wall — a carbon dioxide level that, over time, will cause profound changes.

The risks include melting ice sheets, rising seas, more droughts and heat waves, more flash floods, worse storms, extinction of many plants and animals, depletion of sea life and — perhaps most important — difficulty in producing an adequate supply of food. Many of these changes are taking place at a modest level already, the scientists say, but are expected to intensify.

Reacting to such warnings, President George Bush committed the United States in 1992 to limiting its emissions of greenhouse gases, especially carbon dioxide. Scores of other nations made the same pledge, in a treaty that was long on promises and short on specifics.

But in 1998, when it came time to commit to details in a document known as the Kyoto Protocol, Congress balked. Many countries did ratify the protocol, but it had only a limited effect, and the past decade has seen little additional progress in controlling emissions.

Many countries are reluctant to commit themselves to tough emission limits, fearing that doing so will hurt economic growth. International climate talks in Cancún, Mexico, this month ended with only modest progress. The Obama administration, which came into office pledging to limit emissions in the United States, scaled back its ambitions after climate and energy legislation died in the Senate this year.

Challengers have mounted a vigorous assault on the science of climate change. Polls indicate that the public has grown more doubtful about that science. Some of the Republicans who will take control of the House of Representatives in January have promised to subject climate researchers to a season of new scrutiny.

One of them is Representative Dana Rohrabacher, Republican of California. In a recent Congressional hearing on global warming, he said, "The CO2 levels in the atmosphere are rather undramatic."

But most scientists trained in the physics of the atmosphere have a different reaction to the increase.

"I find it shocking," said Pieter P. Tans, who runs the government monitoring program of which the Mauna Loa Observatory is a part. "We really are in a predicament here, and it's getting worse every year."

As the political debate drags on, the mute gray boxes atop Mauna Loa keep spitting out their numbers, providing a reality check: not only is the carbon dioxide level rising relentlessly, but the pace of that rise is accelerating over time.

"Nature doesn't care how hard we tried," Jeffrey D. Sachs, the Columbia University economist, said at a recent seminar. "Nature cares how high the parts per million mount. This is running away."

A PASSION FOR PRECISION

Perhaps the biggest reason the world learned of the risk of global warming was the unusual personality of a single American scientist.

Charles David Keeling's son Ralph remembers that when he was a child, his family bought a new home in Del Mar, Calif., north of San Diego. His father assigned him the task of edging the lawn. Dr. Keeling insisted that Ralph copy the habits of the previous owner, an Englishman who had taken pride in his garden, cutting a precise two-inch strip between the sidewalk and the grass.

"It took a lot of work to maintain this attractive gap," Ralph Keeling recalled, but he said his father believed "that was just the right way to do it, and if you didn't do that, you were cutting corners. It was a moral breach."

Dr. Keeling was a punctilious man. It was by no means his defining trait — relatives and colleagues described a man who played a brilliant piano, loved hiking mountains and might settle a friendly argument at dinner by pulling an etymological dictionary off the shelf.

But the essence of his scientific legacy was his passion for doing things in a meticulous way. It explains why, even as challengers try to pick apart every other aspect of climate science, his half-century record of carbon dioxide measurements stands unchallenged.

By the 1950s, when Dr. Keeling was completing his scientific training, scientists had been observing the increasing use of fossil fuels and wondering whether carbon dioxide in the air was rising as a result. But nobody had been able to take accurate measurements of the gas.

As a young researcher, Dr. Keeling built instruments and developed techniques that allowed him to achieve great precision in making such measurements. Then he spent the rest of his life applying his approach.

In his earliest measurements of the air, taken in California and other parts of the West in the mid-1950s, he found that the background level for carbon dioxide was about 310 parts per million.

That discovery drew attention in Washington, and Dr. Keeling soon found himself enjoying government backing for his research. He joined the staff of the Scripps Institution of Oceanography, in the La Jolla section of San Diego, under the guidance of an esteemed scientist named Roger Revelle, and began laying plans to measure carbon dioxide around the world.

Some of the most important data came from an analyzer he placed in a government geophysical observatory that had been set up a few years earlier in a remote location: near the top of Mauna Loa, one of the volcanoes that loom over the Big Island of Hawaii.

He quickly made profound discoveries. One was that carbon dioxide oscillated slightly according to the seasons. Dr. Keeling realized the reason: most of the world's land is in the Northern Hemisphere, and plants there were taking up carbon dioxide as they sprouted leaves and grew over the summer, then shedding it as the leaves died and decayed in the winter.

He had discovered that the earth itself was breathing.

A more ominous finding was that each year, the peak level was a little higher than the year before. Carbon dioxide was indeed rising, and quickly. That finding electrified the small community of scientists who understood its implications. Later chemical tests, by Dr. Keeling and others, proved that the increase was due to the combustion of fossil fuels.

The graph showing rising carbon dioxide levels came to be known as the Keeling Curve. Many Americans have never heard of it, but to climatologists, it is the most recognizable emblem of their science, engraved in bronze on a building at Mauna Loa and carved into a wall at the National Academy of Sciences in Washington.

By the late 1960s, a decade after Dr. Keeling began his measurements, the trend of rising carbon dioxide was undeniable, and scientists began to warn of the potential for a big increase in the temperature of the earth.

Dr. Keeling's mentor, Dr. Revelle, moved to Harvard, where he lectured about the problem. Among the students in the 1960s who first saw the Keeling Curve displayed in Dr. Revelle's classroom was a senator's son from Tennessee named Albert Arnold Gore Jr., who marveled at what it could mean for the future of the planet.

Throughout much of his career, Dr. Keeling was cautious about interpreting his own measurements. He left that to other people while he concentrated on creating a record that would withstand scrutiny.

John Chin, a retired technician in Hawaii who worked closely with Dr. Keeling, recently described the painstaking steps he took, at Dr. Keeling's behest, to ensure accuracy. Many hours were required every week just to be certain that the instruments atop Mauna Loa had not drifted out of kilter.

The golden rule was "no hanky-panky," Mr. Chin recalled in an interview in Hilo, Hawaii. Dr. Keeling and his aides scrutinized the records closely, and if workers in Hawaii fell down on the job, Mr. Chin said, they were likely to get a call or letter: "What did you do? What happened that day?"

In later years, as the scientific evidence about climate change grew, Dr. Keeling's interpretations became bolder, and he began to issue warnings. In an essay in 1998, he replied to claims that global warming was a myth, declaring that the real myth was that “natural resources and the ability of the earth’s habitable regions to absorb the impacts of human activities are limitless.”

Still, by the time he died, global warming had not become a major political issue. That changed in 2006, when Mr. Gore’s movie and book, both titled “An Inconvenient Truth,” brought the issue to wider public attention. The Keeling Curve was featured in both.

In 2007, a body appointed by the United Nations declared that the scientific evidence that the earth was warming had become unequivocal, and it added that humans were almost certainly the main cause. Mr. Gore and the panel jointly won the Nobel Peace Prize.

But as action began to seem more likely, the political debate intensified, with fossil-fuel industries mobilizing to fight emission-curbing measures. Climate-change contrarians increased their attack on the science, taking advantage of the Internet to distribute their views outside the usual scientific channels.

In an interview in La Jolla, Dr. Keeling’s widow, Louise, said that if her husband had lived to see the hardening of the political battle lines over climate change, he would have been dismayed.

“He was a registered Republican,” she said. “He just didn’t think of it as a political issue at all.”

THE NUMBERS

Not long ago, standing on a black volcanic plain two miles above the Pacific Ocean, the director of the Mauna Loa Observatory, John E. Barnes, pointed toward a high metal tower.

Samples are taken by hoses that snake to the top of the tower to ensure that only clean air is analyzed, he explained. He described other measures intended to guarantee an accurate record. Then Dr. Barnes, who works for the National Oceanic and Atmospheric Administration, displayed the hourly calculation from one of the analyzers.

It showed the amount of carbon dioxide that morning as 388 parts per million.

After Dr. Keeling had established the importance of carbon dioxide measurements, the government began making its own, in the early 1970s. Today, a NOAA monitoring program and the Scripps Institution of Oceanography program operate in parallel at Mauna Loa and other sites, with each record of measurements serving as a quality check on the other.

The Scripps program is now run by Ralph Keeling, who grew up to become a renowned atmospheric scientist in his own right and then joined the Scripps faculty. He took control of the measurement program after his father's sudden death from a heart attack.

In an interview on the Scripps campus in La Jolla, Ralph Keeling calculated that the carbon dioxide level at Mauna Loa was likely to surpass 400 by May 2014, a sort of odometer moment in mankind's alteration of the atmosphere.

"We're going to race through 400 like we didn't see it go by," Dr. Keeling said.

What do these numbers mean?

The basic physics of the atmosphere, worked out more than a century ago, show that carbon dioxide plays a powerful role in maintaining the earth's climate. Even though the amount in the air is tiny, the gas is so potent at trapping the sun's heat that it effectively works as a one-way blanket, letting visible light in but stopping much of the resulting heat from escaping back to space.

Without any of the gas, the earth would most likely be a frozen wasteland — according to a recent study, its average temperature would be colder by roughly 60 degrees Fahrenheit. But scientists say humanity is now polluting the atmosphere with too much of a good thing.

In recent years, researchers have been able to put the Keeling measurements into a broader context. Bubbles of ancient air trapped by glaciers and ice sheets have been tested, and they show that over the past 800,000 years, the amount of carbon dioxide in the air oscillated between roughly 200 and 300 parts per million. Just before the Industrial Revolution, the level was about 280 parts per million and had been there for several thousand years.

That amount of the gas, in other words, produced the equable climate in which human civilization flourished.

Other studies, covering many millions of years, show a close association between carbon dioxide and the temperature of the earth. The gas seemingly played a major role in amplifying the effects of the ice ages, which were caused by wobbles in the earth's orbit.

The geologic record suggests that as the earth began cooling, the amount of carbon dioxide fell, probably because much of it got locked up in the ocean, and that fall amplified the initial cooling. Conversely, when the orbital wobble caused the earth to begin warming, a great deal of carbon dioxide escaped from the ocean, amplifying the warming.

Richard B. Alley, a climate scientist at Pennsylvania State University, refers to carbon dioxide as the master control knob of the earth's climate. He said that because the wobbles in the earth's orbit were not, by themselves, big enough to cause the large changes of the ice ages, the situation made sense only when the amplification from carbon dioxide was factored in.

"What the ice ages tell us is that our physical understanding of CO2 explains what happened and nothing else does," Dr. Alley said. "The ice ages are a very strong test of whether we've got it right."

When people began burning substantial amounts of coal and oil in the 19th century, the carbon dioxide level began to rise. It is now about 40 percent higher than before the Industrial Revolution, and humans have put half the extra gas into the air since just the late 1970s. Emissions are rising so rapidly that some experts fear that the amount of the gas could double or triple before emissions are brought under control.

The earth's history offers no exact parallel to the human combustion of fossil fuels, so scientists have struggled to calculate the effect.

Their best estimate is that if the amount of carbon dioxide doubles, the temperature of the earth will rise about five or six degrees Fahrenheit. While that may sound small given the daily and seasonal variations in the weather, the number represents an annual global average, and therefore an immense addition of heat to the planet.

The warming would be higher over land, and it would be greatly amplified at the poles, where a considerable amount of ice might melt, raising sea levels. The deep ocean would also absorb a tremendous amount of heat.

Moreover, scientists say that an increase of five or six degrees is a mildly optimistic outlook. They cannot rule out an increase as high as 18 degrees Fahrenheit, which would transform the planet.

Climate-change contrarians do not accept these numbers.

The Internet has given rise to a vocal cadre of challengers who question every aspect of the science — even the physics, worked out in the 19th century, that shows that carbon dioxide traps heat. That is a point so elementary and well-established that demonstrations of it are routinely carried out by high school students.

However, the contrarians who have most influenced Congress are a handful of men trained in atmospheric physics. They generally accept the rising carbon dioxide numbers, they recognize that the increase is caused by human activity, and they acknowledge that the earth is warming in response.

But they doubt that it will warm nearly as much as mainstream scientists say, arguing that the increase is likely to be less than two degrees Fahrenheit, a change they characterize as manageable.

Among the most prominent of these contrarians is Richard Lindzen of the Massachusetts Institute of Technology, who contends that as the earth initially warms, cloud patterns will shift in a way that should help to limit the heat buildup. Most climate scientists contend that little evidence supports this view, but Dr. Lindzen is regularly consulted on Capitol Hill.

"I am quite willing to state," Dr. Lindzen said in a speech this year, "that unprecedented climate catastrophes are not on the horizon, though in several thousand years we may return to an ice age."

THE FUEL OF CIVILIZATION

While the world's governments have largely accepted the science of climate change, their efforts to bring emissions under control are lagging.

The simple reason is that modern civilization is built on burning fossil fuels. Cars, trucks, power plants, steel mills, farms, planes, cement factories, home furnaces — virtually all of them spew carbon dioxide or lesser heat-trapping gases into the atmosphere.

Developed countries, especially the United States, are largely responsible for the buildup that has taken place since the Industrial Revolution. They have begun to make some headway on the problem, reducing the energy they use to produce a given amount of economic output, with some countries even managing to lower their total emissions.

But these modest efforts are being swamped by rising energy use in developing countries like China, India and Brazil. In those lands, economic growth is not simply desirable — it is a moral imperative, to lift more than a third of the human race out of poverty. A recent scientific paper referred to China's surge as "the biggest transformation of human well-being the earth has ever seen."

China's citizens, on average, still use less than a third of the energy per person as Americans. But with 1.3 billion people, four times as many as the United States, China is so large and is growing so quickly that it has surpassed the United States to become the world's largest overall user of energy.

Barring some big breakthrough in clean-energy technology, this rapid growth in developing countries threatens to make the emissions problem unsolvable.

Emissions dropped sharply in Western nations in 2009, during the recession that followed the financial crisis, but that decrease was largely offset by continued growth in the East. And for 2010, global emissions are projected to return to the rapid growth of the past decade, rising more than 3 percent a year.

Many countries have, in principle, embraced the idea of trying to limit global warming to two degrees Celsius, or 3.6 degrees Fahrenheit, feeling that any greater warming would pose unacceptable risks. As best scientists can calculate, that means about one trillion tons of carbon can be burned and the gases released into the atmosphere before emissions need to fall to nearly zero.

"It took 250 years to burn the first half-trillion tons," Myles R. Allen, a leading British climate scientist, said in a briefing. "On current trends, we'll burn the next half-trillion in less than 40."

Unless more serious efforts to convert to a new energy system begin soon, scientists argue, it will be impossible to hit the 3.6-degree target, and the risk will increase that global warming could spiral out of control by century's end.

"We are quickly running out of time," said Josep G. Canadell, an Australian scientist who tracks emissions.

In many countries, the United States and China among them, a conversion of the energy system has begun, with wind turbines and solar panels sprouting across the landscape. But they generate only a tiny fraction of all power, with much of the world's electricity still coming from the combustion of coal, the dirtiest fossil fuel.

With the exception of European countries, few nations have been willing to raise the cost of fossil fuels or set emissions caps as a way to speed the transformation. In the United States, a particular fear has been that a carbon policy will hurt the country's industries as they compete with companies abroad whose governments have adopted no such policy.

As he watches these difficulties, Ralph Keeling contemplates the unbending math of carbon dioxide emissions first documented by his father more than a half-century ago and wonders about the future effects of that increase.

"When I go see things with my children, I let them know they might not be around when they're older," he said. " 'Go enjoy these beautiful forests before they disappear. Go enjoy the glaciers in these parks because they won't be around.' It's basically taking note of what we have, and appreciating it, and saying goodbye to it."

On Dec. 11, another round of international climate negotiations, sponsored by the United Nations, concluded in Cancún. As they have for 18 years running, the gathered nations pledged renewed efforts. But they failed to agree on any binding emission targets.

Late at night, as the delegates were wrapping up in Mexico, the machines atop the volcano in the middle of the Pacific Ocean issued their own silent verdict on the world's efforts.

At midnight Mauna Loa time, the carbon dioxide level hit 390 — and rising.

Temperature Rising: Articles in this series are focusing on the central arguments in the climate debate and examining the evidence for global warming and its consequences.

A Change in Temperature

COLUMN | BY JUSTIN GILLIS | MAY 13, 2013

SINCE 1896, SCIENTISTS have been trying to answer a deceptively simple question: What will happen to the temperature of the earth if the amount of carbon dioxide in the atmosphere doubles?

Some recent scientific papers have made a splash by claiming that the answer might not be as bad as previously feared. This work — if it holds up — offers the tantalizing possibility that climate change might be slow and limited enough that human society could adapt to it without major trauma.

Several scientists say they see reasons to doubt that these lowball estimates will in fact stand up to critical scrutiny, and a wave of papers offering counterarguments is already in the works. "The story is not over," said Chris E. Forest, a climate expert at Pennsylvania State University.

Still, the recent body of evidence — and the political use that climate contrarians are making of it to claim that everything is fine — sheds some light on where we are in our scientific and public understanding of the risks of climate change.

The topic under discussion is a number called "climate sensitivity." Finding this number is the holy grail of climate science, because the stakes are so high: The fate of the earth hangs in the balance.

The first to take a serious stab at it was a Swede named Svante Arrhenius, in the late 19th century. After laborious calculations, he declared that if humans doubled the carbon dioxide in the air by burning fossil fuels, the average temperature of the earth would rise by something like nine degrees Fahrenheit, a whopping figure.

He was on the high side, as it turned out. In 1979, after two decades of meticulous measurements had made it clear that the carbon dioxide level was indeed rising, scientists used computers and a much deeper understanding of the climate to calculate a likely range of warming.

They found that the response to a doubling of carbon dioxide would not be much below three degrees Fahrenheit, nor was it likely to exceed eight degrees.

In the years since, scientists have been pushing and pulling within that range, trying to settle on a most likely value. Most of those who are expert in climatology subscribe to a best-estimate figure of just over five degrees Fahrenheit.

That may not sound like a particularly scary number to many people — after all, we experience temperature variations of 20 or 30 degrees in a single day. But as an average for the entire planet, five degrees is a huge number.

The ocean, covering 70 percent of the surface, helps bring down the average, but the warming is expected to be higher over land, causing weather extremes like heat waves and torrential rains. And the poles will warm even more, so that the increase in the Arctic could exceed 10 or 15 degrees Fahrenheit. That could cause substantial melting of the polar ice sheets, ultimately flooding the world's major coastal cities.

What's new is that several recent papers have offered best estimates for climate sensitivity that are below four degrees Fahrenheit, rather than the previous best estimate of just above five degrees, and they have also suggested that the highest estimates are pretty implausible.

Notice that these recent calculations fall well within the long-accepted range — just on the lower end of it. But the papers have caused considerable excitement among climate-change contrarians.

It is not that they actually agree with the new numbers, mind you. They have long pushed implausibly low estimates of climate sensitivity, below two degrees Fahrenheit in some cases. But they appear to be calculating that any paper with a lowball number is a step in their direction.

James Annan, a mainstream climate scientist working at a Japanese institute, offers a best estimate of four and a half degrees Fahrenheit. When he wrote recently that he thought some of the highest temperature projections could be rejected, skeptics could not contain their enthusiasm.

"That is what we call a landmark change of course — by one of climatology's most renowned warmist scientists," declared a blogger named Pierre L. Gosselin. "If even Annan can see it, then the writing is truly emblazoned on the wall."

But does this sort of claim — that we can all breathe a sigh of relief about climate change — really hold up?

Dr. Annan said in an e-mail that the Intergovernmental Panel on Climate Change, a mainstream body that periodically summarizes climate science, should be bolder about ruling out extreme temperature scenarios, but he still believes global warming is a sufficient threat to warrant changes in human behavior.

He noted that climate skeptics "are desperate to claim that the I.P.C.C. is being unreasonably alarmist, but on the other hand they don't really want to agree with me either, because my views are close enough to the mainstream as to be unacceptable to them." He added that he finds it "amusing to watch their gyrations as they try to square the circle."

It will certainly be good news if these recent papers stand up to critical scrutiny, something that will take at least a year or two to figure out. But the need for additional scientific vetting before we accept the lower numbers is not the biggest flaw in the contrarian argument.

Remember, the climate sensitivity number, whatever it turns out to be, applies to a doubling of carbon dioxide.

Given how weak the political response to climate change has been, there is no reason to think that human society is going to stop there. Some experts think the level of the heat-trapping gas could triple or even quadruple before emissions are reined in. Only last week the level of carbon dioxide passed a milestone of 400 parts per million at the flagship monitoring station atop Mauna Loa, in Hawaii, evidence that efforts to control emissions are failing.

Even if climate sensitivity turns out to be on the low end of the range, total emissions may wind up being so excessive as to drive the earth toward dangerous temperature increases.

So if the recent science stands up to critical examination, it could indeed turn into a ray of hope — but only if it is then followed by a broad new push to get the combustion of fossil fuels under control.

JUSTIN GILLIS writes about our changing climate in the By Degrees column for The New York Times.

Why Should We Trust Climate Models?

BY JUSTIN GILLIS | SEPT. 23, 2014

Q. *Despite decades of work with climate models, science has failed to produce a single model with any predictive value. Why should we trust them? Clearly they don't work. — Asked by KeMa*

A. THANKS FOR THE QUESTION, but your statement is dubious. Climate models have proven to have quite a bit of predictive value. In the first paper on global warming, published in 1896, the Swedish scientist Svante Arrhenius constructed a simple set of equations predicting that the earth would warm from the carbon dioxide humans were pumping into the atmosphere; it took 80 years to be sure he was right, but he was. In the 1960s, the first truly elaborate climate models predicted that the Arctic would warm faster than the earth as a whole; that turned out to be correct. Early climate models predicted that the higher levels of the atmosphere would cool as the lower levels warmed; that turned out to be correct. In the 1980s and 1990s, climate models predicted that we would start to get more intense rains as global warming proceeded; that turned out to be correct. There are many, many examples like this.

What is true, however, is that modern climate models are not very good at predicting short-term climatic variations that are influenced by natural cycles like El Niño and La Niña; they were not designed to do this. For that reason, the models did not foresee the slowdown in global warming that has occurred over the past 15 years or so.

Scientists would certainly like to improve the models to the point that they can make such short-term predictions, and some of them think they are closing in on the goal. All climate scientists acknowledge imperfections in the models, but far from hiding this, they talk about it endlessly in huge scientific conferences that anyone can attend.

A Prophet of Doom Was Right About the Climate

OPINION | BY JUSTIN GILLIS | JUNE 23, 2018

THE NIGHT BEFORE the day that would make him famous, James E. Hansen listened to a baseball game on the radio. But his mind kept wandering: What would he say to Congress the next day to convey that humans were endangering the planet?

He had long been trying to raise the alarm without success, and so had other scientists. But then, on June 23, 1988 — 30 years ago Saturday — a Colorado senator named Tim Wirth convened yet another hearing on the topic. Dr. Hansen was one of several scientists on the witness list.

Few people had ever heard of him, nor of the obscure NASA unit that he headed. He and a small group of colleagues studied the Earth's climate, working in a suite of offices above the Manhattan diner that "Seinfeld" would later make famous.

He had conducted rigorous studies of historical temperatures, concluding that the planet was warming sharply. He had helped to pioneer computer modeling of the climate, and the results predicted further warming if people kept pouring greenhouse gases into the atmosphere.

June 23 turned out be a blistering day in Washington, and much of the nation was suffering through a drought and heat wave. Dr. Hansen took his seat in a Capitol Hill hearing room and laid out the scientific facts as best he understood them.

He had thought up a good line the night before, during the Yankees game, but in the moment he forgot to deliver it. When the hearing ended, though, reporters surrounded him, and he remembered.

"It is time to stop waffling so much," he said, "and say that the evidence is pretty strong that the greenhouse effect is here."

His near certainty that human emissions were already altering the climate caught the attention of a sweltering nation, catapulting Dr.

Hansen to overnight fame. That year, 1988, would go on to be the hottest in a global temperature record stretching back to the 19th century.

With the perspective of three decades, it is fair to ask: How right was his forecast?

The question defies a simple answer. In 1988, Dr. Hansen had to offer a prognostication not just about how the Earth would respond to greenhouse gases, but also about how much of those gases humans would choose to inject into the air.

He did what any cautious forecaster would do: He offered low, medium and high scenarios. The warming over the past 30 years has indeed fallen well within his upper and lower bounds.

One of Dr. Hansen's scenarios, Scenario B, has turned out to be a reasonably close match for fossil-fuel emissions as they actually occurred. Yet we now know Scenario B predicted too much global warming, by something like 30 percent.

Two reasons for that stand out. One is that Dr. Hansen had assumed a continued increase in certain refrigerant gases that warm the climate. Those gases were ultimately brought under control by a global treaty, the Montreal Protocol — proof that scientific warnings, if taken seriously, can be acted upon at a worldwide scale.

The bigger problem was that the computers he was using in the 1980s could not operate fast enough to give a realistic picture of the upper atmosphere; as a result, his model was most likely overestimating the Earth's sensitivity to emissions. In the years since, computer modeling of the climate, though hardly perfect, has improved.

So while his temperature forecast was not flawless, in a larger sense, Dr. Hansen's 1988 warning has turned out to be entirely on target. As emissions have soared, the planet has warmed relentlessly, just as he said it would; 1988 is not even in the top 20 warmest years now. Every year of this century has been hotter.

The ocean is rising, as Dr. Hansen predicted, and the pace seems to be accelerating. The great ice sheets in Greenland and Antarctica are dumping ever-rising volumes of water into the sea. Coastal

AGATA NOWICKA

flooding is increasing rapidly in the United States. The Arctic Ocean ice cap has shrunk drastically.

If his warning in 1988 had been met with a national policy to reduce emissions, other countries might have followed, and the world would be in much better shape.

But within a few years after he raised the alarm, fossil-fuel interests and libertarian ideologues began financing a campaign of lies about climate research. The issue bogged down in Congress, and to this day that body has taken no action remotely commensurate with the threat.

Dr. Hansen retired from NASA in 2013, but at age 77, he feels his work is not done. Today, from an office at Columbia University, he spends his time fighting the government he once served. He is an expert witness for a lawsuit that young people have filed in Oregon against the federal government, contending that its failure to tackle climate change is a threat to their constitutional rights of life and liberty.

His granddaughter, Sophie Kivlehan, is one of the plaintiffs in the case, which has gotten much farther than many legal experts thought it would. The case may go to trial later this year.

Prophets of impending calamity are rarely thanked for their efforts, especially when they turn out to be right. But Dr. Hansen did receive a form of thanks recently, sharing half a of a $1.3 million prize for his attempts to warn the public about the risks of climate change.

The congressional failure to respond to his warning might be seen now as a harbinger of the political crisis that has since engulfed the United States. How can Congress tackle global warming if it lacks the capacity to solve far smaller problems?

Lately, Dr. Hansen has been thinking about the connection between the political crisis and the climate crisis. He is a strong proponent of a new system of voting, called ranked choice, that has been adopted in many other countries and a few parts of the United States, with the goal of recreating a political center.

"It's very hard to see us fixing the climate," Dr. Hansen said, "until we fix our democracy."

JUSTIN GILLIS is a former New York Times environmental reporter and a contributing opinion writer.

CHAPTER 2

Reading the Signs of Climate Change

The signs of climate change are most visible on increasingly hot days and in regular reports of the melting polar regions. But scientists and the general public are starting to notice other indicators. These include storms and droughts whose heightened severity makes them likely indicators of a changing climate. Many such droughts are exacerbating the migrant crisis as well. Researchers, tracking conditions' likelihood using statistical analysis, are beginning to identify these signs with confidence. Meanwhile, the damages are only getting worse.

Pace of Ocean Acidification Has No Parallel in 300 Million Years, Paper Says

BY JUSTIN GILLIS | MARCH 2, 2012

A NEW SCIENTIFIC PAPER suggests that the ocean is acidifying at a rate that is many times faster than at any time in the last 300 million years. The change is occurring so rapidly that it raises "the possibility that we are entering an unknown territory of marine ecosystem change," said the paper, published this week in the journal Science.

The new study, led by Bärbel Hönisch, a Columbia University paleoceanographer, does not present much new scientific evidence on the issue. Instead, it is a careful analysis of the existing evidence from decades of research on the earth's geologic history.

That history features some fast releases of carbon dioxide into the atmosphere that in some ways resemble the current trend of release to which humans greatly contribute by burning fossil fuels today. Those historical releases warmed the planet just as it is warming now. Because much of the extra carbon dioxide released into the air gets deposited in the ocean as a mild acid, past events also caused the ocean to turn more acidic.

But as scientists have long known, and the new paper reiterates, those previous releases were usually much slower than the one occurring now. (While the present-day release of carbon dioxide is slow on a human time scale, it is essentially instantaneous on a geologic time scale.)

So scientists have been struggling to figure out whether any of those past events can serve as good analogues for the present, providing us with some sense of the environmental and biological changes that may be in store for the earth.

Slow though they may have been, some of the past events caused profound planetary change. One of them, a stupendous series of volcanic eruptions in what is now Siberia starting about 250 million years ago, put so much carbon dioxide into the air over the course of a million years that it apparently wiped out most species on earth. Life took tens of millions of years to recover.

A more interesting analogy, however, may be the Paleocene-Eocene Thermal Maximum 55 million years ago, in which a pulse of carbon dioxide from an unknown source entered the atmosphere over several thousand years. That event produced immense environmental changes and some extinctions of life in the sea, but research suggests it did not lead to mass extinctions on land.

It did produce a rapid proliferation of new species as land animals adjusted to the environmental shifts. Our own lineage, the primates, apparently blossomed during that event, filling new ecological niches.

The new paper suggests that ocean acidification is now unfolding at at least 10 times the rate that it was during the Paleocene-Eocene

Thermal Maximum. Thus, the scientists conclude that no past event is likely to serve as a perfect analogue for the human release of carbon dioxide, which "stands out as capable of driving a combination and magnitude of ocean geochemical changes potentially unparalleled" in 300 million years, the paper says.

Patrizia Ziveri, a researcher from the Autonomous University of Barcelona who took part in the study, said the findings underlined the need for rapid policy action. "Considering the effects we detect through fossil records, there is no doubt that we must tackle the problem at its roots as soon as possible, adopting measures to immediately reduce our CO2 emissions into the atmosphere," she said in a statement.

Paris Accord Considers Climate Change as a Factor in Mass Migration

BY SEWELL CHAN | DEC. 12, 2015

LE BOURGET, FRANCE — The two-week United Nations climate conference outside Paris that drew to an end on Saturday focused on many of the physical dangers associated with climate change: extreme weather, severe drought, the warming of oceans, rain forest destruction and disruptions to the food supply.

But global warming has already had another effect — the large-scale displacement of people — that has been an ominous, politically sensitive undercurrent in the talks and side events here.

Scientists have said that climate change can indirectly lead to migration by setting off violent conflicts. Scholars have made this connection since at least 2007, when they cited climate change as a reason for the war in Darfur, Sudan.

A drought that lasted from 2006 to 2011 in much of Syria has been cited as a factor in the long-running civil war there, fueling a mass migration to Turkey, Lebanon and Jordan, but also to Europe, Canada and, in small measure, the United States.

Europe, in particular, is experiencing the largest influx of migrants since World War II — Germany alone has already taken in nearly a million this year. Jean-Claude Juncker, the president of the European Commission, told world leaders on Nov. 30 that climate change could "destabilize entire regions and start massive forced migrations and conflicts over natural resources."

The Paris climate accord, adopted on Saturday, calls for developing recommendations "to avert, minimize and address displacement related to the adverse impacts of climate change" — an explicit acknowledgment of the dangers of migration that some of the poorest of the 195 countries involved in the talks had sought to include in the text.

From 2008 to 2014, an average of 26.4 million people were displaced each year by floods, storms, earthquakes and other natural disasters, according to a report released in July by the Internal Displacement Monitoring Center, part of the Norwegian Refugee Council. Most moved within their countries.

"Climate-related displacement is not a future phenomenon," said Marine Franck, who works on climate change and migration for the United Nations high commissioner for refugees. "It is a reality; it is already a global concern."

William Lacy Swing, a retired American ambassador who now leads the International Organization for Migration, said that climate change was adding to a "perfect storm" of "unprecedented human mobility," a result of the quadrupling of the world's population over the last century and wars, conflicts and persecution that have displaced a record 60 million people.

He said that migration had to be viewed "not as a problem to be solved, but a human reality that has to be managed."

The United Nations Convention to Combat Desertification — like the climate talks, it grew out of the Earth Summit in Rio de Janeiro in 1992 — and the British Defense Ministry recently cited a 2009 report estimating that 135 million people are at risk of displacement because desertification, the drying out of once-fertile land, will reduce drinking-water supplies and lower coral yields. The problem is most pronounced across a band of Africa, from the Sahel in the west to the Horn of Africa in the east.

By 2020, some 60 million people could move from the desertified areas of sub-Saharan Africa toward North Africa and Europe, the report found; by 2050, about 200 million people may be permanently displaced.

The report was prepared by a research and advocacy organization led by Kofi Annan, the former United Nations secretary general, that shut down in 2010; some of the report's findings have been disputed. Indeed, the numbers are so staggering that Jan Egeland, the head of the Norwegian Refugee Council, who with Mr. Swing and others spoke at a panel discussion here, took pains to point out that the vast majority of migration worldwide takes place in the developing world.

As early as 1990, a report by the Intergovernmental Panel on Climate Change warned, "The greatest single impact of climate change could be on human migration."

It was not until 20 years later, at the 2010 United Nations climate conference in Cancun, Mexico, that countries formally agreed that "climate change-induced migration, displacement and relocation" were among the challenges the world faced in adapting to a warmer planet.

In 2012, the Norwegian and Swiss governments established a research entity, the Nansen Initiative, which found that "a serious legal gap exists with regard to cross-border movements in the context of disasters and the effects of climate change." The initiative has held consultations in four particularly vulnerable regions — Central America, the Horn of Africa, Southeast Asia and the islands of the South Pacific — and plans to recommend a "protection agenda" that may include standards of treatment.

People forced to leave their homes because of climate change are not easily classified under existing human rights, refugee or asylum law. In July, a New Zealand court dismissed a landmark case brought by a man from Kiribati, Ioane Teitiota, who had sought to have his family classified as "climate change refugees." They were deported in September.

SOMINI SENGUPTA contributed reporting from the United Nations.

Looking, Quickly, for the Fingerprints of Climate Change

BY HENRY FOUNTAIN | AUG. 1, 2016

WHEN DAYS OF heavy rain in late May caused deadly river flooding in France and Germany, Geert Jan van Oldenborgh got to work.

Dr. van Oldenborgh is not an emergency responder or a disaster manager, but a climate researcher with the Royal Netherlands Meteorological Institute. With several colleagues around the world, he took on the task of answering a question about the floods, one that arises these days whenever extreme weather occurs: Is climate change to blame?

For years, most meteorologists and climate scientists would answer that question with a disclaimer, one that was repeated so often it became like a mantra: It is not possible to attribute individual weather events like storms, heat waves or droughts to climate change.

But increasingly over the past decade, researchers have been trying to do just that, aided by better computer models, more weather data and, above all, improved understanding of the science of a changing climate.

Attribution studies, as they are called, can take many months, in large part because of the time needed to run computer models. Now scientists like Dr. van Oldenborgh, who is part of a group called World Weather Attribution, are trying to do such studies much more quickly, as close to the event as possible.

"Scientific teams are taking on the challenge of doing this kind of analysis rather rapidly," said Peter A. Stott, who leads the climate attribution group at the Met Office, Britain's weather agency.

The goal is to get sound scientific analysis to the public to help counter misinformation, deliberate or otherwise, about an event.

"It's worthwhile to give the best scientific evidence at the time, rather than not saying anything and letting others say things that are

ROBERT STOLARIK FOR THE NEW YORK TIMES

Flooding in the streets of Queens, N.Y. Researchers are beginning to measure the links between current weather events and long-term climate change.

not related to what really happened," said Friederike Otto, a researcher at the University of Oxford who works with Dr. van Oldenborgh.

In the case of the European floods, World Weather Attribution, which is coordinated by Climate Central, a climate-change research organization based in Princeton, N.J., released a report less than two weeks after the Seine and other rivers overtopped their banks. The group concluded that climate change had made the French flooding more likely, but could not draw a conclusion about the flooding in Germany.

"In the French case, we had five almost independent measures, and they all agreed," Dr. van Oldenborgh said. "With Germany, we only had two, and they disagreed."

Climate scientists have said for decades that global warming should lead to an increase in extreme weather like heat waves and droughts. Because more water evaporates from the oceans and warmer air holds more moisture, climate change should also lead to more intense and frequent storms.

Studies have shown that these effects are occurring on a broad scale. The National Climate Assessment, for instance, notes that heavy downpours have increased across most of the United States in the last 25 years.

But analyzing individual events is problematic, largely for two reasons: Weather is naturally variable, even without climate change, and global warming may be only one of several factors influencing a particular event. Since reliable data is required, studies are also less likely to be undertaken for events in countries that lack much data-collecting infrastructure, or where governments do not share data widely.

Dr. Stott, of the Met Office, was the lead researcher for an early attribution study, a 2004 paper in Nature that linked a deadly 2003 heat wave in Europe to human-caused increases in greenhouse gas concentrations in the atmosphere. Since then the pace of such studies has increased; last year, a publication of the American Meteorological Society, edited by Dr. Stott and others, had 32 studies of 28 events, covering all seven continents.

Not all of them found a link to climate change. A study of water shortages in southeastern Brazil during a dry period in late 2014 and early 2015, for instance, found that the shortages were most likely driven by increasing population and water use than by climate change.

But other analyses — of a 2013 heat wave in Argentina, extreme rainfall in the Cévennes Mountains in France in 2014 and an extremely hot spring in South Korea that year — found a connection.

David W. Titley, a professor of meteorology at Pennsylvania State University who was chairman of a National Academies committee that looked at developments in the field of climate-change attribution, said that at this point studies of heat waves and other extreme-temperature events appeared to produce the most reliable assessments.

Studies of extreme rainfall are considered less reliable in finding links to climate change, and studies of events like wildfires and severe thunderstorms even less reliable.

Still, Dr. Titley said, such studies are worth doing, as long as certain conditions are met.

"There are very legitimate reasons why people want to do this rapidly," he said. "But they need to state very clearly what the assumptions are, what the methods are, what the confidence level is," he said.

"This is still not like predicting what time is sunrise for New York City tomorrow," he added.

Attribution studies usually involve running climate models many times over. Because no model is a perfect representation of reality, varying them slightly for each run and averaging the results give scientists more confidence in their accuracy.

One set of runs simulates the climate as it actually is, incorporating the effects of greenhouse gas emissions, while the other set simulates the climate as if that human influence had never happened.

Researchers then compare historical data, as well as data from the actual event, such as rainfall or temperatures, to the different model results to assess any climate-change impact. The analysis is usually given in terms of probabilities, or increased or reduced likelihood, that climate change had an effect.

Rather than running models after an event, researchers like Dr. van Oldenborgh and Dr. Otto shorten the process by using models that have already been run.

"The only way we can do this rapid attribution is by precooking everything that we can," Dr. van Oldenborgh said.

The process starts with emails among members of the group, usually followed by a Skype session to discuss whether a specific event is worthy of study. (One group member is in Australia, so arranging a conference call can be tricky.)

At least one member of the group also has to have time to do the work. "You have to put everything aside for a week" said Dr. van

Oldenborgh, who did most of the work on the flood analysis because the other researchers were at a conference.

A crucial part of the task is to define the event — what happened and what meteorological variable is best to study it. In the case of the European study, the models the researchers used simulate rainfall, not flooding. So they used rainfall as the variable.

"We can't look at flooding, only extreme precipitation," Dr. Otto said. "Where did it rain, how much did it rain?"

But they also consulted with a hydrologist who understood how rainfall affects the river systems involved, Dr. van Oldenborgh said, "just to make sure we didn't do something stupid."

Among the models they use is one that Dr. Otto's group at Oxford, the Environmental Change Institute, runs regularly, using the personal computers of a large number of volunteers, a project called climateprediction.net. While the model is a global one, regional results can be extracted and used for the rapid analysis.

Dr. van Oldenborgh said that the European flooding analysis was a good illustration of the need for transparency, because while the researchers were confident in their findings for France, they acknowledged that they could not draw conclusions regarding Germany.

"We have to make sure we're open-minded enough to conclude that, although we've spent a lot of time on this, we can't conclude anything," he said.

When the initial analysis is completed and released to the public — Climate Central helps with the communication process — the work is still not over. Although the hope is that such studies will eventually become so routine that there is no need to publish the findings in a scientific journal, for now, at least, a research paper has to be written and submitted for peer review. So for the flooding analysis, that was another week's work for Dr. van Oldenborgh.

"The worst problem is that you're pretty tired and want to take a break and put your feet up," he said. "We haven't got a solution for that yet."

Scientists Link Hurricane Harvey's Record Rainfall to Climate Change

BY HENRY FOUNTAIN | DEC. 13, 2017

NEW ORLEANS — Climate change made the torrential rains that flooded Houston after Hurricane Harvey last summer much worse, scientists reported Wednesday.

Two research groups found that the record rainfall as Harvey stalled over Texas in late August, which totaled more than 50 inches in some areas, was as much as 38 percent higher than would be expected in a world that was not warming.

While many scientists had said at the time that Harvey was probably affected by climate change, because warmer air holds more moisture, the size of the increase surprised some.

"The amount of precipitation increase is worse than I expected," said Michael J. Wehner, a senior scientist at Lawrence Berkeley National Laboratory and an author of a paper on his group's findings, which included the 38 percent figure. Based on how much the world has warmed, Dr. Wehner said, before the analysis he had expected an increase of only about 6 or 7 percent.

The other study, by an international coalition of scientists known as World Weather Attribution, found that Harvey's rainfall was 15 percent higher than would be expected without climate change. Geert Jan van Oldenborgh of the Royal Netherlands Meteorological Institute and the lead author of the second study, said that climate change also made such an extreme rainstorm much more likely.

"The probability of such an event has increased by roughly a factor of three," he said. While the likelihood of a Harvey-like storm was perhaps once in every 3,000 years in the past, he said, now it's once every 1,000 years or so — which means that in any given year, there is 0.1 percent chance of a similar storm occurring along the Gulf Coast.

Harvey developed in the Gulf of Mexico and made landfall near Corpus Christi, Tex., as a strong Category 4 hurricane. By the time it reached Houston it had weakened to a tropical storm, but it moved slowly over the region, rotating and picking up more moisture from the Gulf of Mexico. Thousands of homes and businesses in the region were flooded and more than 80 people died.

David W. Titley, a meteorologist at Pennsylvania State University who was not involved with either study, said the research showed that "while a storm of Harvey's strength is still rare, it's not as rare as it once was.

"Communities all along the Gulf Coast need to adapt to a world where the heaviest rains are more than we have ever seen," he added.

Antonia Sebastian, a researcher at Rice University and a co-author of the World Weather Attribution paper, said that Harvey was a much larger event than governments and developers normally plan and build for.

"What we see from this study is that the flood hazard zone isn't stationary," Dr. Sebastian said. "Precipitation is changing, and that's changing the boundaries. That should be considered."

In August, a week before Hurricane Harvey, President Trump rolled back an Obama-era executive order that included climate change and sea-level rise in federal flood risk standards.

Both studies were released during the annual meeting here of the American Geophysical Union, a large gathering of leading climate researchers and other earth scientists.

The studies only looked at the impact of climate change on rainfall, not whether warming affected Harvey's formation or strength. Those issues remain a subject of much debate among scientists, with some researchers suggesting that strong hurricanes — category 4 and above — will become more frequent as the world continues to warm.

Teasing out the influence of climate change on hurricanes remains extremely problematic, Dr. van Oldenborgh said.

"The effect of climate change on hurricanes is horribly complicated," he said. "We're working on it, but it's very difficult."

But the more limited analysis, determining the influence of warming on the rainfall of a huge storm like Harvey "turns out to be a solvable problem," Dr. van Oldenborgh said.

The studies are the latest in a series of analyses that search for the fingerprints of climate change on individual weather events like storms or heat waves. Despite overwhelming evidence that the climate is changing overall because of greenhouse gases emissions, for years most scientists had said it was extremely difficult to link warming to specific events.

That has now changed, with studies in recent years that found that climate change affected Australian heat waves in 2013, downpours in Louisiana in 2016, floods in France that same year and many other events. In some cases — German floods around the same time as the French ones, for example — studies have been inconclusive or found no link to climate change.

Although there were some differences, the two current studies employed the same basic approach — making use of actual data from the storm, and comparing two sets of climate models, those that take into account existing conditions, in which rising carbon dioxide has warmed the planet, and those that assume CO2 emissions had never happened and the climate is as it was more than a century ago.

Dr. Titley, who served as chairman of a National Academies committee that looked at developments in the field of climate-change attribution, said that both studies were "carefully done and combine observations with the latest simulation techniques."

Dr. Wehner's study, in particular, raises the issue of whether climate change might have contributed to the slow motion of Harvey, a subject that Dr. Titley said was worthy of further research.

Why So Cold? Climate Change May Be Part of the Answer

BY HENRY FOUNTAIN | JAN. 3, 2018

AS BITTER COLD continues to grip much of North America and helps spawn the fierce storm along the East Coast, the question arises: What's the influence of climate change?

Some scientists studying the connection between climate change and cold spells, which occur when cold Arctic air dips south, say that they may be related. But the importance of the relationship is not fully clear yet.

The Arctic is not as cold as it used to be — the region is warming faster than any other — and studies suggest that this warming is weakening the jet stream, which ordinarily acts like a giant lasso, corralling cold air around the pole.

"There's a lot of agreement that the Arctic plays a role, it's just not known exactly how much," said Marlene Kretschmer, a researcher at the Potsdam Institute for Climate Impact Research in Germany. "It's a very complex system."

The reason a direct connection between cold weather and global warming is still up for debate, scientists say, is that there are many other factors involved. Ocean temperatures in the tropics, soil moisture, snow cover, even the long-term natural variability of large ocean systems all can influence the jet stream.

"I think everyone would agree that potentially the warming Arctic could have impacts on the lower latitudes," said Rick Thoman, climate services manager with the National Weather Service in Fairbanks, Alaska. "But the exact connection on the climate scale is an area of active research."

Much of the Northern Hemisphere is cold this time of year (it's winter, after all). Cold snaps have occurred throughout history — certainly long before industrialization resulted in large emissions of greenhouse gases. And as with any single weather event, it's difficult to directly attribute the influence of climate change to a particular cold spell.

But scientists have been puzzled by data that at first seems counterintuitive: Despite an undeniable overall year-round warming trend, winters in North America and Europe have trended cooler over the past quarter-century.

"We're trying to understand these dynamic processes that lead to cold winters," Ms. Kretschmer said.

She is the lead author of a study published last fall that looked at four decades of climate data and concluded that the jet stream — usually referred to as the polar vortex this time of year — is weakening more frequently and staying weaker for longer periods of time. That allows cold air to escape the Arctic and move to lower latitudes. But the study focused on Europe and Russia.

"The changes in very persistent weak states actually contributed to cold outbreaks in Eurasia," Ms. Kretschmer said. "The bigger question is how this is related to climate change."

Timo Vihma, head of the polar meteorology and climatology group at the Finnish Meteorological Institute, explained that warmer air in the Arctic reduces the temperature difference between it and lower latitudes and weakens the polar vortex.

"When we have a weak temperature gradient between the Arctic and mid-latitudes, the result is weaker winds," he said.

Ordinarily the jet stream is straight, blowing from west to east. When it becomes weaker, Dr. Vihma said, it can become wavy, "more like a big snake around the Northern Hemisphere."

The weaker winds are more susceptible to disturbances, such as a zone of high pressure that can force colder air southward. These "blocking" high-pressure zones are often what creates a severe cold spell that lingers for several days or longer.

The current cold snap has been in place for more than a week, and the cold air on Wednesday was moving east and colliding with a mass of warmer air from the Atlantic Ocean. That created a storm known as a "bomb cyclone."

In a bomb cyclone, the temperature difference between the two air masses leads to a steep and rapid — meteorologists often use the term "explosive" — drop in atmospheric pressure. The air starts to move and, aided by the earth's rotation, begins to rotate. The swirling air can bring high winds and a lot of precipitation, often in the form of snow.

That could happen this time — depending on the track of the storm, parts of the Northeast were expecting heavy snow. But one impact of the storm is even more clear: After it eventually moves off to the north and west, it should draw even more cold polar air into the eastern half of the United States, continuing the big chill.

‘Like a Terror Movie’: How Climate Change Will Cause More Simultaneous Disasters

BY JOHN SCHWARTZ | NOV. 19, 2018

GLOBAL WARMING IS posing such wide-ranging risks to humanity, involving so many types of phenomena, that by the end of this century some parts of the world could face as many as six climate-related crises at the same time, researchers say.

This chilling prospect is described in a paper published Monday in Nature Climate Change, a respected academic journal, that shows the effects of climate change across a broad spectrum of problems, including heat waves, wildfires, sea level rise, hurricanes, flooding, drought and shortages of clean water.

Such problems are already coming in combination, said the lead author, Camilo Mora of the University of Hawaii at Manoa. He noted that Florida had recently experienced extreme drought, record high temperatures and wildfires — and also Hurricane Michael, the powerful Category 4 storm that slammed into the Panhandle last month. Similarly, California is suffering through the worst wildfires the state has ever seen, as well as drought, extreme heat waves and degraded air quality that threatens the health of residents.

Things will get worse, the authors wrote. The paper projects future trends and suggests that, by 2100, unless humanity takes forceful action to curb the greenhouse gas emissions that drive climate change, some tropical coastal areas of the planet, like the Atlantic coast of South and Central America, could be hit by as many as six crises at a time.

That prospect is “like a terror movie that is real,” Dr. Mora said.

The authors include a list of caveats about the research: Since it is a review of papers, it will reflect some of the potential biases of science in this area, which include the possibility that scientists might focus on negative effects more than positive ones; there is also a margin of uncertainty involved in discerning the imprint of climate change from natural variability.

New York can expect to be hit by four climate crises at a time by 2100 if carbon emissions continue at their current pace, the study says, but if emissions are cut significantly that number could be reduced to one. The troubled regions of the coastal tropics could see their number of concurrent hazards reduced from six to three.

The paper explores the ways that climate change intensifies hazards and describes the interconnected nature of such crises. Greenhouse gas emissions, by warming the atmosphere, can enhance drought in places that are normally dry, "ripening conditions for wildfires and heat waves," the researchers say. In wetter areas, a warmer atmosphere retains more moisture and strengthens downpours, while higher sea levels increase storm surge and warmer ocean waters can contribute to the overall destructiveness of storms.

In a scientific world marked by specialization and siloed research, this multidisciplinary effort by 23 authors reviewed more than 3,000 papers on various effects of climate change. The authors determined 467 ways in which those changes in climate affect human physical and mental health, food security, water availability, infrastructure and other facets of life on Earth.

The paper concludes that traditional research into one element of climate change and its effects can miss the bigger picture of interrelation and risk.

Climate change also has different ramifications for the world's haves and have-nots, the authors found: "The largest losses of human life during extreme climatic events occurred in developing nations, whereas developed nations commonly face a high economic burden of damages and requirements for adaptation."

People are not generally attuned to dealing with problems like climate change, Dr. Mora said. "We as humans don't feel the pain of people who are far away or far into the future," he said. "We normally care

ERIC THAYER FOR THE NEW YORK TIMES

A search-and-rescue team looking for human remains in the aftermath of the recent Camp Fire in Paradise, Calif. The state is also suffering from drought, extreme heat waves and degraded air quality.

about people who are close to us or that are impacting us, or things that will happen tomorrow."

And so, he said, people tend to look at events far in the future and tell themselves, "We can deal with these things later, we have more pressing problems now." But, he added, this research "documented how bad this already is."

The paper includes an interactive map of the various hazards under different emissions scenarios for any location in the world, produced by Esri, which develops geographic information systems. "We see that climate change is literally redrawing the lines on the map, and revealing the threats that our world faces at every level," said Dawn Wright, the company's chief scientist.

Michael E. Mann, a climate scientist at Pennsylvania State University who was not involved in the paper, said it underscored the urgency for action to curb the effects of climate change and showed that "the costs of inaction greatly outweigh the costs of taking action."

Dr. Mann published a recent paper suggesting that climate change effects on the jet stream are contributing to a range of extreme summer weather events, such as heat waves in North America, Europe and Asia, wildfires in California and flooding in Japan. The new study, he said, dovetails with that research, and "is, if anything, overly conservative" — that is, it may underestimate the threats and costs associated with human-caused climate change.

A co-author of the new paper, Kerry Emanuel of the Massachusetts Institute of Technology, hailed its interdisciplinary approach. "There's more than one kind of risk out there," he said, but scientists tend to focus on their area of research. "Nations, societies in general, have to deal with multiple hazards, and it's important to put the whole picture together."

Like military leaders developing the capability to fight wars on more than one front, governments have to be ready to face more than one climate crisis at a time, Dr. Emanuel said.

Dr. Mora said he had considered writing a book or a movie that would reflect the frightening results of the research. His working title, which describes how dire the situation is for humanity, is unprintable here. His alternate title, he said, is "We Told You So."

JOHN SCHWARTZ is part of the climate team. Since joining The Times in 2000, he has covered science, law, technology, the space program and more, and has written for almost every section.

Warming in Arctic Raises Fears of a 'Rapid Unraveling' of the Region

BY JOHN SCHWARTZ AND HENRY FOUNTAIN | DEC. 11, 2018

PERSISTENT WARMING IN the Arctic is pushing the region into "uncharted territory" and increasingly affecting the continental United States, scientists said Tuesday.

"We're seeing this continued increase of warmth pervading across the entire Arctic system," said Emily Osborne, an official with the National Oceanic and Atmospheric Administration, who presented the agency's annual assessment of the state of the region, the "Arctic Report Card."

The Arctic has been warmer over the last five years than at any time since records began in 1900, the report found, and the region is warming at twice the rate as the rest of the planet.

Dr. Osborne, the lead editor of the report and manager of NOAA's Arctic Research Program, said the Arctic was undergoing its "most unprecedented transition in human history."

In 2018, "warming air and ocean temperatures continued to drive broad long-term change across the polar region, pushing the Arctic into uncharted territory," she said at a meeting of the American Geophysical Union in Washington.

The rising air temperatures are having profound effects on sea ice, and on life on land and in the ocean, scientists said. The impacts can be felt far beyond the region, especially since the changing Arctic climate may be influencing extreme weather events around the world.

The new edition of the report does not present a radical break with past installments, but it shows that troublesome trends wrought by climate change are intensifying. Air temperatures in the Arctic in 2018 will be the second-warmest ever recorded, the report said, behind only 2016.

Susan M. Natali, an Arctic scientist at Woods Hole Research Center in Massachusetts who was not involved in the research, said the report was another warning going unheeded. "Every time you see a report, things get worse, and we're still not taking any action," she said. "It adds support that these changes are happening, that they are observable."

The warmer Arctic air causes the jet stream to become "sluggish and unusually wavy," the researchers said. That has possible connections to extreme weather events elsewhere on the globe, including last winter's severe storms in the United States and a bitter cold spell in Europe known as the "Beast From the East."

The jet stream normally acts as a kind of atmospheric spinning lasso that encircles and contains the cold air near the pole; a weaker, wavering jet stream can allow Arctic blasts to travel south in winter and can stall weather systems in the summer, among other effects.

"On the East Coast of the United States where the other part of the wave comes down," Dr. Osborne said, "you have these Arctic air temperatures that are surging over into the lower latitudes and causing these crazy winter storms."

The rapid warming in the upper north, known as Arctic amplification, is tied to many factors, including the simple fact that snow and ice reflect a lot of sunlight, while open water, which is darker, absorbs more heat. As sea ice melts, less ice and more open water create a "feedback loop" of more melting that leads to progressively less ice and more open water.

And as Arctic waters become increasingly ice-free, there are commercial and geopolitical implications: New shipping routes may open, and rivalries with other countries, including Russia, are intensifying.

The federal government has issued the report card since 2006. It has continued to do so under the Trump administration, which has approved other scientific reports about global warming and the human greenhouse gas emissions that cause it, despite President Trump's rejection of climate science.

Over all, "the effects of persistent Arctic warming continue to mount," the new report said. "Continued warming of the Arctic atmosphere and ocean are driving broad change in the environmental system in predicted and, also, unexpected ways."

Some of the findings in the research, provided by 81 scientists in 12 countries, included:

• The wintertime maximum extent of sea ice in the region, in March of this year, was the second lowest in 39 years of record keeping.

• Ice that persists year after year, forming thick layers, is disappearing from the Arctic. This is important because the very old ice tends to resist melting; without it, melting accelerates. Old ice made up less than 1 percent of the Arctic ice pack this year, a decline of 95 percent over the last 33 years.

• Donald K. Perovich, a sea-ice expert at Dartmouth College who contributed to the report, said the "big story" for ice this year was in the Bering Sea, off western Alaska, where the extent of sea ice reached a record low for virtually the entire winter. During two weeks in February, normally a time when sea ice grows, the Bering Sea lost an area of ice the size of Idaho, Dr. Perovich said.

• The lack of ice and surge of warmth coincides with rapid expansion of algae species in the Arctic Ocean, associated with harmful blooms that can poison marine life and people who eat the contaminated seafood. The northward shift of the algae "means that the Arctic is now vulnerable to species introductions into local communities and ecosystems that have little to no prior exposure to this phenomenon," the report said.

• Reindeer and caribou populations have declined 56 percent in the past two decades, dropping to 2.1 million from 4.7 million. Scientists monitoring 22 herds found that two of them were at peak numbers without declines, but five populations had declined more than 90 percent "and show no sign of recovery."

• Tiny bits of ocean plastic, which can be ingested by marine life, are proliferating at the top of the planet. "Concentrations in the remote Arctic Ocean are higher than all other ocean basins in the world," the report said. The microplastics are also showing up in Arctic sea ice. Scientists have found samples of cellulose acetate, used in making cigarette filters, and particles of plastics used in bottle caps and packaging material.

"The report card continues to document a rapid unraveling of the Arctic," said Rafe Pomerance, chairman of Arctic 21, a network of organizations focused on educating policymakers and others on Arctic climate change. "The signals of decline are so powerful and the consequences so great that they demand far more urgency from all governments to reduce emissions."

The report was issued as delegates from nearly 200 countries were meeting in Poland for the latest round of climate talks stemming from the Paris Agreement, the landmark climate accord that was designed to reduce planet-warming emissions.

Mr. Trump has vowed to withdraw from the agreement. At the talks, the United States joined with Saudi Arabia, Kuwait and Russia in refusing to endorse a major report to the conference on the effects of climate change around the world.

At a news conference Tuesday announcing the findings of the Arctic report, Tim Gallaudet, a retired Navy admiral who is the acting NOAA administrator, was asked if he or any other senior NOAA officials had ever briefed Mr. Trump on climate change or the changes in the Arctic.

"The simple answer is no," he said.

JOHN SCHWARTZ is part of the climate team. Since joining The Times in 2000, he has covered science, law, technology, the space program and more, and has written for almost every section.

HENRY FOUNTAIN covers climate change, with a focus on the innovations that will be needed to overcome it. He is the author of "The Great Quake," a book about the 1964 Alaskan earthquake.

Humans Are Speeding Extinction and Altering the Natural World at an 'Unprecedented' Pace

BY BRAD PLUMER | MAY 6, 2019

WASHINGTON — Humans are transforming Earth's natural landscapes so dramatically that as many as one million plant and animal species are now at risk of extinction, posing a dire threat to ecosystems that people all over the world depend on for their survival, a sweeping new United Nations assessment has concluded.

The 1,500-page report, compiled by hundreds of international experts and based on thousands of scientific studies, is the most exhaustive look yet at the decline in biodiversity across the globe and the dangers that creates for human civilization. A summary of its findings, which was approved by representatives from the United States and 131 other countries, was released Monday in Paris. The full report is set to be published this year.

Its conclusions are stark. In most major land habitats, from the savannas of Africa to the rain forests of South America, the average abundance of native plant and animal life has fallen by 20 percent or more, mainly over the past century. With the human population passing 7 billion, activities like farming, logging, poaching, fishing and mining are altering the natural world at a rate "unprecedented in human history."

At the same time, a new threat has emerged: Global warming has become a major driver of wildlife decline, the assessment found, by shifting or shrinking the local climates that many mammals, birds, insects, fish and plants evolved to survive in. When combined with the other ways humans are damaging the environment, climate change is now pushing a growing number of species, such as the Bengal tiger, closer to extinction.

LALO DE ALMEIDA FOR THE NEW YORK TIMES

Cattle grazing on a tract of illegally cleared Amazon forest in Pará State, Brazil. In most major land habitats, the average abundance of native plant and animal life has fallen by 20 percent or more, mainly over the past century.

As a result, biodiversity loss is projected to accelerate through 2050, particularly in the tropics, unless countries drastically step up their conservation efforts.

The report is not the first to paint a grim portrait of Earth's ecosystems. But it goes further by detailing how closely human well-being is intertwined with the fate of other species.

"For a long time, people just thought of biodiversity as saving nature for its own sake," said Robert Watson, chair of the Intergovernmental Science-Policy Platform on Biodiversity and Ecosystem Services, which conducted the assessment at the request of national governments. "But this report makes clear the links between biodiversity and nature and things like food security and clean water in both rich and poor countries."

A previous report by the group had estimated that, in the Americas, nature provides some $24 trillion of non-monetized benefits to humans each year. The Amazon rain forest absorbs immense quantities of carbon dioxide and helps slow the pace of global warming. Wetlands purify drinking water. Coral reefs sustain tourism and fisheries in the Caribbean. Exotic tropical plants form the basis of a variety of medicines.

But as these natural landscapes wither and become less biologically rich, the services they can provide to humans have been dwindling.

Humans are producing more food than ever, but land degradation is already harming agricultural productivity on 23 percent of the planet's land area, the new report said. The decline of wild bees and other insects that help pollinate fruits and vegetables is putting up to $577 billion in annual crop production at risk. The loss of mangrove forests and coral reefs along coasts could expose up to 300 million people to increased risk of flooding.

The authors note that the devastation of nature has become so severe that piecemeal efforts to protect individual species or to set up wildlife refuges will no longer be sufficient. Instead, they call for "transformative changes" that include curbing wasteful consumption, slimming down agriculture's environmental footprint and cracking down on illegal logging and fishing.

"It's no longer enough to focus just on environmental policy," said Sandra M. Díaz, a lead author of the study and an ecologist at the National University of Córdoba in Argentina. "We need to build biodiversity considerations into trade and infrastructure decisions, the way that health or human rights are built into every aspect of social and economic decision-making."

Scientists have cataloged only a fraction of living creatures, some 1.3 million; the report estimates there may be as many as 8 million plant and animal species on the planet, most of them insects. Since 1500, at least 680 species have blinked out of existence, including the Pinta giant tortoise of the Galápagos Islands and the Guam flying fox.

Though outside experts cautioned it could be difficult to make precise forecasts, the report warns of a looming extinction crisis, with extinction rates currently tens to hundreds of times higher than they have been in the past 10 million years.

"Human actions threaten more species with global extinction now than ever before," the report concludes, estimating that "around 1 million species already face extinction, many within decades, unless action is taken."

Unless nations step up their efforts to protect what natural habitats are left, they could witness the disappearance of 40 percent of amphibian species, one-third of marine mammals and one-third of reef-forming corals. More than 500,000 land species, the report said, do not have enough natural habitat left to ensure their long-term survival.

Over the past 50 years, global biodiversity loss has primarily been driven by activities like the clearing of forests for farmland, the expansion of roads and cities, logging, hunting, overfishing, water pollution and the transport of invasive species around the globe.

In Indonesia, the replacement of rain forest with palm oil plantations has ravaged the habitat of critically endangered orangutans and Sumatran tigers. In Mozambique, ivory poachers helped kill off nearly 7,000 elephants between 2009 and 2011 alone. In Argentina and Chile, the introduction of the North American beaver in the 1940s has devastated native trees (though it has also helped other species thrive, including the Magellanic woodpecker).

All told, three-quarters of the world's land area has been significantly altered by people, the report found, and 85 percent of the world's wetlands have vanished since the 18th century.

And with humans continuing to burn fossil fuels for energy, global warming is expected to compound the damage. Roughly 5 percent of species worldwide are threatened with climate-related extinction if global average temperatures rise 2 degrees Celsius above preindustrial levels, the report concluded. (The world has already warmed 1 degree.)

"If climate change were the only problem we were facing, a lot of species could probably move and adapt," Richard Pearson, an ecologist at the University College of London, said. "But when populations are already small and losing genetic diversity, when natural landscapes are already fragmented, when plants and animals can't move to find newly suitable habitats, then we have a real threat on our hands."

The dwindling number of species will not just make the world a less colorful or wondrous place, the report noted. It also poses risks to people.

Today, humans are relying on significantly fewer varieties of plants and animals to produce food. Of the 6,190 domesticated mammal breeds used in agriculture, more than 559 have gone extinct and 1,000 more are threatened. That means the food system is becoming less resilient against pests and diseases. And it could become harder in the future to breed new, hardier crops and livestock to cope with the extreme heat and drought that climate change will bring.

"Most of nature's contributions are not fully replaceable," the report said. Biodiversity loss "can permanently reduce future options, such as wild species that might be domesticated as new crops and be used for genetic improvement."

The report does contain glimmers of hope. When governments have acted forcefully to protect threatened species, such as the Arabian oryx or the Seychelles magpie robin, they have managed to fend off extinction in many cases. And nations have protected more than 15 percent of the world's land and 7 percent of its oceans by setting up nature reserves and wilderness areas.

Still, only a fraction of the most important areas for biodiversity have been protected, and many nature reserves poorly enforce prohibitions against poaching, logging or illegal fishing. Climate change could also undermine existing wildlife refuges by shifting the geographic ranges of species that currently live within them.

So, in addition to advocating the expansion of protected areas, the authors outline a vast array of changes aimed at limiting the drivers of biodiversity loss.

Farmers and ranchers would have to adopt new techniques to grow more food on less land. Consumers in wealthy countries would have to waste less food and become more efficient in their use of natural resources. Governments around the world would have to strengthen and enforce environmental laws, cracking down on illegal logging and fishing and reducing the flow of heavy metals and untreated wastewater into the environment.

The authors also note that efforts to limit global warming will be critical, although they caution that the development of biofuels to reduce emissions could end up harming biodiversity by further destroying forests.

None of this will be easy, especially since many developing countries face pressure to exploit their natural resources as they try to lift themselves out of poverty. But, by detailing the benefits that nature can provide to people, and by trying to quantify what is lost when biodiversity plummets, the scientists behind the assessment are hoping to help governments strike a more careful balance between economic development and conservation.

"You can't just tell leaders in Africa that there can't be any development and that we should turn the whole continent into a national park," said Emma Archer, who led the group's earlier assessment of biodiversity in Africa. "But we can show that there are trade-offs, that if you don't take into account the value that nature provides, then ultimately human well-being will be compromised."

In the next two years, diplomats from around the world will gather for several meetings under the Convention on Biological Diversity, a global treaty, to discuss how they can step up their efforts at conservation. Yet even in the new report's most optimistic scenario, through 2050 the world's nations would only slow the decline of biodiversity — not stop it.

"At this point," said Jake Rice, a fisheries scientist who led an earlier report on biodiversity in the Americas, "our options are all about damage control."

BRAD PLUMER is a reporter covering climate change, energy policy and other environmental issues for The Times's climate team.

CHAPTER 3

Fossil Fuels, Polluter Nations and Other Causes of Warming

The driver of climate change is the dependence of industrialized nations on fossil fuels, for everything from power to transportation to food production. Among these nations, China and the United States are the largest polluters. But most industrialized nations struggle to reduce carbon emissions, as ambitious targets face unexpected setbacks. For still-developing countries such as China and India, there is the unique challenge of raising standards of living in a sustainable manner. Alongside these challenges, tropical deforestation has surged, threatening additional warming.

A Shrinking Window for Burning Fossil Fuel

BY JAMES KANTER | MAY 11, 2009

A RECENT ISSUE of the journal Nature carried two scientific reports that underline the challenge for governments — and for fossil fuel reliant countries and companies — if the world agrees to keep greenhouse gas emissions within limits deemed safe by scientists.

Looking at the broad period beginning roughly from the industrial revolution to the year 2500, one of the reports argued the need to limit cumulative carbon dioxide to about 1 trillion metric tons, on the basis that the accumulation of long-lived greenhouse gases, like CO2, mainly determines maximum projected warming.

But with more than that amount of carbon dioxide already released since industrial times, "it may well turn out that we can only afford to release less than the same again, possibly much less, with many times that amount in fossil-fuel reserves remaining underground," according to a commentary by the authors of the reports published by Nature online on April 30.

The commentary warns that "having taken 250 years to burn the first half-trillion tonnes of carbon, we look set, on current trends, to burn the next half trillion in less than 40."

The commentary also notes that "descendants in the second half of this century, knowing much more about climate change and its impacts than we do, may decide that they need to intervene actively to reduce atmospheric CO2 concentrations."

Missing Its Own Goals, Germany Renews Effort to Cut Carbon Emissions

BY MELISSA EDDY | DEC. 3, 2014

BERLIN — Germany has fallen behind in its ambitious goals for reducing carbon emissions. It is burning more coal than at any point since 1990. And German companies are complaining that the nation's energy policies are hurting their ability to compete globally.

But on Wednesday, Chancellor Angela Merkel's government said it was redoubling its efforts, proposing new measures to help it reach the emissions-reduction target for 2020 it set seven years ago when it undertook an aggressive effort to combat climate change.

The new plan was unveiled by a country eager to retain a leadership position in international talks to address the threat from global warming. The plan underscored Ms. Merkel's commitment despite the problems it has caused her at home.

The plan calls on Germans to cut an additional 62 million to 78 million tons of carbon dioxide emissions — the annual output of about seven million German households. That would triple emission reductions from current levels, spreading the cuts across sectors like agriculture and automobiles.

The program, which would be established by laws to be passed by Parliament, rests on improved energy efficiency, with 3 billion euros, or $3.7 billion, in tax breaks and other incentives.

Roughly a third of the cuts are to come from the power industry, even as coal-fired plants continue to play an essential role.

Germany's predicament reflects the difficulty faced by modern economies in reducing carbon as an energy source. But polls show that most Germans favor reducing emissions.

"If we want to keep our promise, we need to close this gap, and that is what we are doing," Barbara Hendricks, Ms. Merkel's environment minister, said at a news conference on Wednesday.

Ms. Hendricks is to present Germany's position next week in Lima, Peru, to leaders who are working to create the basis for a new global agreement on emissions reductions ahead of a world summit in Paris next year. Last month, China and the United States, the world's two biggest polluters, announced plans to lower carbon emissions.

The World Meteorological Organization said Wednesday that 2014 was on track to be the warmest year on record. "This is an important message for negotiators so that they know that decisions have to be taken quickly," Michel Jarraud, the organization's secretary general, said in Geneva. He added that the evidence linking human-generated carbon emissions to climate change was much stronger than it was 20 years ago, and a "lack of knowledge is no longer an excuse for inaction."

Ms. Merkel made her debut on the international stage as Germany's environment minister by marshaling support for the Kyoto agreement in 1997, and has made it clear that she wants Germany to remain at the forefront of efforts to combat climate change.

She helped rally the European Union's 28 leaders around the issue in October, and intends to use her country's turn at the Group of 7 summit meeting next year to push for a Paris accord.

But the German union that represents workers in the mining, chemical and energy industries warned Berlin that the latest round of cuts could affect jobs in the country's coal-rich and industrial regions.

Ahead of Wednesday's announcement, members of the union had gathered thousands of signatures demanding German leaders provide "affordable electricity and good jobs." Sigmar Gabriel, the minister for economic affairs and energy, has pledged to give utilities free rein to decide where and how they make further emissions cuts, with an eye to unions' close ties to his center-left Social Democratic Party.

Last year, 45 percent of Germany's power came from hard coal and the soft brown coal known as lignite, the highest level since 2007, according to data from AG Energiebilanzen, a group of energy lobbying firms and economic research institutes. That compares with 25 percent of energy from renewable resources.

Since Germany began shutting down its 17 nuclear power plants, a drive that gained speed after the 2011 nuclear disaster in Fukushima, Japan, it has increasingly depended on coal-fired plants for the flow of power needed by the country's large industrial base. Its dependence on coal is the highest in nearly 25 years, when many of East Germany's worst-polluting factories and plants were shut down after reunification with the West.

Soft coal, the dirtiest fossil fuel to burn, is Germany's cheapest and most abundant natural resource. Hundreds of thousands of jobs depend on the lignite mines and the power industry they support, and workers are growing increasingly fearful that Germany's energy revolution will cost them their livelihoods.

Many of Germany's leading industries, like chemicals and aluminum, are based in the coal-rich Rhine region, contributing to Germany's post-World War II economic strength. But in recent years, energy-intensive companies have been looking abroad to expand their businesses.

Frank Löllgen, the head of the trade union's western North Rhine region, said the trend worried the 107,000 workers he represented. The union has been urging the government to remember energy's role in the economy's continuing success in recent years.

"We already are on the edge of what is possible," Mr. Löllgen said in an interview at his Düsseldorf office. "Is it worth it if we as a country succeed in reaching our targets in reducing carbon emissions, but sacrifice good jobs and our industrial base?"

NICK CUMMING-BRUCE contributed reporting from Geneva.

Food Waste Is Becoming Serious Economic and Environmental Issue, Report Says

BY RON NIXON | FEB. 25, 2015

WASHINGTON — With millions of households across the country struggling to have enough to eat, and millions of tons of food being tossed in the garbage, food waste is increasingly being seen as a serious environmental and economic issue.

A report released Wednesday shows that about 60 million metric tons of food is wasted a year in the United States, with an estimated value of $162 billion. About 32 million metric tons of it end up in municipal landfills, at a cost of about $1.5 billion a year to local governments.

The problem is not limited to the United States.

The report estimates that a third of all the food produced in the world is never consumed, and the total cost of that food waste could be as high as $400 billion a year. Reducing food waste from 20 to 50 percent globally could save $120 billion to $300 billion a year by 2030, the report found.

"Food waste is a global issue, and tackling it is a priority," said Richard Swannell, director of sustainable food systems at the Waste and Resources Action Program, or Wrap, an antiwaste organization in Britain that compiled the new report. "The difficulty is often in knowing where to start and how to make the biggest economic and environmental savings."

The food discarded by retailers and consumers in the most developed countries would be more than enough to feed all of the world's 870 million hungry people, according to the Food and Agriculture Organization of the United Nations.

But it is not just those countries that have problems with food waste. The report showed that it is also an issue in African countries like South Africa.

The problem is expected to grow worse as the world's population increases, the report found. By 2030, when the global middle class expands, consumer food waste will cost $600 billion a year, unless actions are taken to reduce the waste, according to the report.

Food waste is not only a social cost, but it contributes to growing environmental problems like climate change, experts say, with the production of food consuming vast quantities of water, fertilizer and land. The fuel that is burned to process, refrigerate and transport it also adds to the environmental cost.

Most food waste is thrown away in landfills, where it decomposes and emits methane, a potent greenhouse gas. Globally, it creates 3.3 billion metric tons of greenhouse gases annually, about 7 percent of the total emissions, according to the report.

The United Nations agency points out that methane gas from the world's landfills are surpassed in emissions by only China and the United States.

"Seven percent is not the largest contributor of greenhouse gasses, but it's not an insignificant amount," said Helen Mountford, the director of economics at the World Resources Institute. "But this is one area — reducing food waste — where we can make a difference."

Over the last several years, some cities and counties in the United States, including New York City, have started programs to tackle the issue. Hennepin County, Minn., the state's most populous county, provides grants from $10,000 to $50,000 to local business and nonprofits to help recycle food products or turn them into compost.

"There is still a lot in the waste stream," said Paul Kroening, supervising environmentalist at Hennepin County Environmental Services. "We are just scratching the surface."

A coalition of food industry trade groups, the Food Waste Reduction Alliance, has also increased effort to combat food waste. Meghan Stasz, the director of sustainability for the Grocery Manufacturers Association, a member of the alliance, said the group was working with supermarket chains to reduce waste by clarifying expiration dates and selling smaller portions of food.

Ms. Stasz said the group was also getting its members to donate more food and make changes in manufacturing processes to reduce the amount of wasted food. One member, the giant food company ConAgra, changed the way it placed dough in shell for its pot pies and saved 235 tons of dough in a year.

Mr. Swannell, of the antiwaste group Wrap, applauded those efforts, but said more still needed to be done.

"Awareness of food waste has risen, but we need to do more to tie that awareness to actions on the ground," he said. "We need to find better ways to deal with food waste, but we need to prevent it in the first place."

India Is Caught in a Climate Change Quandary

COLUMN | BY EDUARDO PORTER | NOV. 10, 2015

SPARE A THOUGHT for poor India.

India is home to 30 percent of the world's poorest, those living on less than $1.90 a day. Of the 1.3 billion Indians, 304 million do not have access to electricity; 92 million have no access to safe drinking water.

And India is going to be hammered by climate change.

The livelihoods of 600 million Indians are threatened by the expected disruption of the southwest monsoon from July to September, which accounts for 70 percent of India's rainfall. India's rivers depend on the health of thousands of Himalayan glaciers at risk of melting because of a warming climate, while 150 million people are at risk from storm surges associated with rising sea levels.

A lot of damage is already inevitable, a consequence of the emissions of heat-trapping greenhouse gases by richer countries. So, many Indians ask, Why must we pay more? On what grounds can India be asked to temper its use of energy to limit its emissions of greenhouse gases like carbon dioxide?

"Today, I see the carbon space occupied by the developed world," Prakash Javadekar, the environment minister, said in an interview with The Associated Press in September. "We are asking the developed world to vacate the carbon space to accommodate us. That carbon space demand is climate justice."

The successful resolution of this confrontation of priorities does not matter just for India's sake. The tension between economic development and the imperative to curb greenhouse gas emissions remains the central challenge of the diplomatic effort to muster a coalition of rich and poor countries to combat climate change.

The United Nations expects India's population to reach 1.5 billion by 2030, bigger than China's. If over the next 15 years it follows anything like the fossil-fuel-heavy path out of poverty that China took over the last 15, it could blow any chance the world has of preventing a disaster.

A critical question for anyone with a stake in preventing a climatic catastrophe is how to conceive and finance a development path for 1.5 billion Indians that prevents this outcome.

Scientists and environmentalists, executives and government diplomats packing their bags to attend the climate summit meeting in Paris, starting Nov. 30, must keep the challenge in mind.

After so many failed rounds of diplomacy, everyone involved is eager to declare the coming meeting a success. So far, 129 countries accounting for nearly 90 percent of greenhouse gas emissions have submitted plans to contribute to the cause.

While the progress is undoubtedly real, the central challenge remains unresolved. Countries are not being asked to make legally binding commitments to reduce their greenhouse gas emissions. They will show up, instead, with "Intended Nationally Determined Contributions" to the mitigation effort.

Advanced countries will offer absolute cuts in carbon emissions. But the less developed are expected only to reduce their emissions intensity — a measure of the carbon dioxide released to produce a certain amount of economic activity — in a recognition that their energy consumption still has a long way to grow.

The new approach was necessary to achieve any progress. But it required putting the tough questions aside. Nearly as populous as China, yet way behind in terms of economic development, India presents one of the tougher ones.

By most accounts, the world's greenhouse gas emissions must be brought close to zero by the end of the century, at the latest. This constrains everyone.

For instance, a recent report by the World Bank argues that economies like China and India must totally decarbonize their electricity supply around midcentury and achieve negative emissions from then on, using carbon capture technologies and vastly increased forests, to suck excessive carbon out of the atmosphere.

To put it mildly, that is going to be a challenge.

Jairam Ramesh, who was minister of the environment under the previous prime minister, Manmohan Singh, argues that India must continue to grow at 7.5 to 8 percent a year for the next 15 years.

To power this growth, India's electricity consumption — which accounts for over half its greenhouse gas emissions — would rise 6 to 7 percent a year. Even under the most ambitious goals for nuclear power and renewable energy, more than half of this power is expected to come from coal, the dirtiest fuel. "By 2030 India's coal consumption could triple or quadruple," Mr. Ramesh told me.

India has come up with a mitigation contribution plan for the Paris meeting. It aims to get 40 percent of its electricity from non-fossil fuels by 2030 and to reduce its emissions intensity by 33 to 35 percent from 2005 to 2030. It also offers to vastly increase its forest cover.

The plan, however, pointedly notes that India's energy consumption amounts to only 0.6 metric tons of oil equivalent per person, about a third of the world average. It explains that "no country in the world" has ever achieved the development level of today's advanced nations without consuming at least four tons.

"India has a lot to do to provide a dignified life to its population and meet their rightful aspirations," it states.

Some analysts say there is a way to thread the needle. Development can be decoupled from carbon emissions, the World Bank insists.

Moreover, economists at the World Bank argued in a separate report released last Sunday that emissions reduction policies could be structured to benefit the poor in the next 15 years — for instance by using revenue from carbon taxes to pay for social insurance.

"The goals are extremely ambitious; only a minority of scenarios get us there," said Stephane Hallegatte, who led the study. "But they are achievable."

Under the right set of policies, the World Bank projects, even the most disruptive climate change would add only three million people to India's extreme poor in 2030. Bad choices, by contrast, would add 42 million to that number.

Some in India seem convinced by the logic. "Our traditional defensive stance has simply not been in the enlightened national interest," Mr. Ramesh argued in an address last year to the National Institute of Advanced Studies in Bangalore.

"India must view the era of the green economy not as a threat to its developmental plans," he said. "Instead, it must be viewed as an opportunity to build and demonstrate technological capability to the world."

And yet, there is still a significant risk that India will say no to the West's climate change agenda. "It plays hugely well domestically," Mr. Ramesh told me. "One should never discount that possibility."

EDUARDO PORTER writes the Economic Scene column for The New York Times. Economic Scene explores the world's most urgent economic encounters.

Calls for Shipping and Aviation to Do More to Cut Emissions

BY HENRY FOUNTAIN | APRIL 16, 2016

EVEN THOUGH COMMERCIAL AVIATION and ocean shipping are significant sources of greenhouse gas emissions, they were excluded from the Paris climate treaty, to be signed by more than 100 countries this week at the United Nations in New York.

Now governments and advocacy groups are pressuring these industries to take stronger steps to curb pollution.

A coalition of European, North African and South Pacific nations is lobbying the International Maritime Organization, the United Nations agency that oversees shipping, to start discussing an emissions-reduction commitment at a meeting in London that will begin Monday.

"We need to do something and go beyond what we already have, and set some very specific targets," said François Martel, the secretary general of the Pacific Islands Development Forum. The forum's members include the Marshall Islands and the Solomon Islands, two of six nations that have made a proposal, expected to be taken up at the meeting, that shipping contribute a "fair share" to reducing emissions.

Another United Nations agency, the International Civil Aviation Organization, has for years been considering a market-based strategy in which airlines could purchase "offsets," or emissions reductions from renewable energy or conservation projects, to cover at least some international flights.

Advocacy groups are pressuring the agency to adopt as strict a system as possible when it meets for its triennial assembly in Montreal this fall.

"If we're going to have offsets, then they actually have to deliver the tons of reductions they say they will," said Bill Hemmings, the director of aviation and shipping at Transport & Environment, an environmental group based in Brussels.

Nigel Purvis, the chief executive of Climate Advisers, a consulting group in Washington, said airlines were likely to increase spending significantly on offsets from forest conservation projects.

"Airlines know this sector and are ready to play," he said.

While some previous forest projects have been criticized for not delivering the reductions that were claimed, "now we have new rules about how to do forests in a way that as we scale up we maintain integrity," Mr. Purvis added.

Aviation and shipping each contribute a little more than 2 percent of annual worldwide human-produced emissions of carbon dioxide. Together that is more than the emissions from Japan, the world's fifth-largest emitter.

Both industries are expected to grow over the next few decades, and their percentages of worldwide emissions may increase significantly as emissions are reduced elsewhere. Environmental groups say steps the industries have already taken, including regulations to reduce emissions from new aircraft and ships, will not help much because they are tied to baselines for improvement that are too low.

Yet after being included in initial drafts of the climate treaty, a paragraph on limiting or reducing emissions from the two industries was eliminated from the final version, which was agreed upon in Paris in mid-December.

The treaty commits nations to setting emissions-reduction targets, with a goal of keeping global warming "well below" a target of 2 degrees Celsius, or 3.6 degrees Fahrenheit, above preindustrial levels.

Experts cite several reasons that aviation and shipping were not in the treaty, including a desire to keep the text as concise as possible to improve the chances of reaching an agreement. The issue also would have exacerbated disputes about the responsibilities of developed versus developing nations that could have threatened the overall accord, they said.

Industry representatives and environmental groups alike say that despite the lack of any mention in the treaty, there is still momentum for action on emissions by both industries.

Simon Bennett, the director of policy and external relations for the International Chamber of Shipping, an industry group, said that there was a “misunderstanding” about the Paris accord and that “somehow that means shipping has escaped.”

“That isn’t the case,” Mr. Bennett said. The chamber has filed its own proposal for the International Maritime Organization meeting; it uses language other than “fair share” but still calls for emissions-reductions targets.

But there are disagreements between the shipping industry and environmental advocates about the best ways to cut emissions. The industry generally favors a global fuel tax over carbon offsets, and notes that most ships have already reduced their emissions and that there is a maritime organization program in place, the Energy Efficiency Design Index, to reduce emissions from new ones.

Environmental groups, however, argue that the efficiency index program’s improvement standards are too low, and that most ships built in the last several years already meet the standards for 2020.

“They need to come up with more stringent targets,” Mr. Hemmings of Transport & Environment said.

The aviation industry also points out that it is not relying solely on so-called market-based measures like offsets to reduce emissions.

“The global offsetting scheme is just one aspect of the sector’s climate action, albeit a crucial one,” said Michael Gill, the executive director of the Air Transport Action Group, an industry organization.

Like shipping, aviation has adopted efficiency standards. The International Civil Aviation Organization approved them in February, and will limit emissions from jets built after 2023 from current designs, and from new models introduced after 2028.

Critics say that those standards are weak, and that most advanced jets being built already meet them. That makes adopting tough market-based measures more important than ever, they say.

"The level of the CO2 efficiency standard for new aircraft, set in February, was disappointing in its ambition," said Kat Watts, a global climate policy adviser with Carbon Market Watch, in Brussels. With aviation left out of the Paris treaty, she added, the International Civil Aviation Organization "was handed the baton for climate action for international aviation."

"Whether they run with, or drop, that baton will be decided in this October's assembly," she added.

In the United States, the Environmental Protection Agency has begun what is expected to be a yearslong process to develop emissions rules for aircraft, and has said the rules would be at least as strict as the international organization's standards.

But environmental groups have argued that the E.P.A.'s rules must be far more stringent. Last week, several groups, including Friends of the Earth, sued the environmental agency in an effort to compel it to move faster to develop the rules.

'Irrational' Coal Plants May Hamper China's Climate Change Efforts

BY EDWARD WONG | FEB. 7, 2017

YINING, CHINA — When scientists and environmental scholars scan the grim industrial landscape of China, a certain coal plant near the rugged Kazakhstan border stands out.

On the outside, it looks like any other modern energy plant — shiny metal towers loom over the grassy grounds, and workers in hard hats stroll the campus. But in those towers, a rare and contentious process is underway, spewing an alarming amount of carbon dioxide, the main greenhouse gas accelerating climate change.

The plant and others like it undermine China's aim of being a global leader on efforts to limit climate change.

The plant, in the country's far west, converts coal to synthetic natural gas. The process, called coal-to-gas or coal gasification, has been criticized by Chinese and foreign scholars and policy makers. For one thing, it is relatively expensive. It also requires enormous amounts of water, which exacerbates the chronic water crisis in northern China. And worst of all, critics say, it emits more carbon dioxide than traditional methods of energy production, even other coal-based ways.

"It is extremely irrational to develop coal-to-gas technology," Li Junfeng, a climate change and energy adviser to the government, wrote in 2015 in China Energy News, a publication managed by People's Daily, the Communist Party newspaper. He added that coal-to-gas was "unfit to become a national strategy."

Despite such denunciations and a continuing policy debate, at least four such plants have begun operating in China in the past four years, pushed by local governments and state-owned enterprises in coal-rich regions. Dozens more have been under consideration.

No other country is considering building coal-to-gas plants on this scale.

STEPHEN CROWLEY/THE NEW YORK TIMES

President Xi Jinping of China with President Obama at a ceremony in Hangzhou announcing the formal adoption of the Paris climate agreement in 2016.

The technology's emerging use reveals the challenges China faces in reducing coal's central role in the economy, particularly in the north, which has some of the world's largest coal deposits.

Under the Obama administration, United States officials working on climate change tracked China's use of this technology and expressed concern to Chinese counterparts, including the environmental protection minister, Chen Jining. That will most likely change under President Trump, who has called climate change a "hoax" and threatened to renounce the Paris agreement, a global accord to limit carbon emissions.

In January, President Xi Jinping said at the World Economic Forum in Davos, Switzerland, that all countries should remain committed to the Paris agreement because "this is a responsibility we must shoulder for the sake of our future generations."

The Chinese government is trying to rein in local officials and companies in northern China that undermine central policy when acting out of economic self-interest. Too often, the actions of local governments contradict attempts by central policy makers to wean Chinese industries off coal and move toward a less carbon-heavy economy.

"With the economy and market conditions, we really don't think this is a good choice," said Gan Yiwei, a climate and energy campaigner at Greenpeace East Asia, based in Beijing, who has researched coal-to-gas plants.

The plant here in the frontier region of Xinjiang is a pioneer effort by coal proponents. It was opened by a local subsidiary of the China Kingho Energy Group, the country's largest private coal mining company, and is outside the city of Yining, whose population is a mix of ethnic Han, Uighur, Kazakh and other groups.

Ethnic tensions in the area sometimes explode into protests or violence, and police officers in riot gear stand at barricaded posts on many of Yining's street corners. The tensions are worsened by the large presence of the energy industry here — Xinjiang has vast coal and oil deposits, and ethnic minorities are aware that state-owned energy companies exploiting local resources are run by ethnic Han and employ mostly Han workers.

In 2013, Kingho opened the coal-to-gas plant as a pilot project closely watched by central-level officials. Xinjiang has the largest coal reserves in China. The same year, a similar pilot project opened in Inner Mongolia, another region brimming with coal.

After the market for coal slumped in 2012, local governments in northern China desperately began looking for new uses for it. Officials turned to two industries in which coal is converted to other material: coal-to-gas and coal-to-chemicals. Officials have expanded both industries across northern China, despite their detrimental environmental impact.

Two Duke University researchers published an essay in 2013 recommending that Chinese officials cancel the coal-to-gas programs. Some central government officials understood that there were risks to adopting these technologies. The National Development and

Reform Commission, the government's economic planning agency, held a forum in January 2014 for experts to discuss the effects of coal-to-gas on climate change. There have been no public reports on the conference.

In July 2014, Greenpeace East Asia issued a report that said governments and companies across China had plans to operate 50 coal-to-gas plants. That would produce an estimated 1.1 billion tons of carbon dioxide per year — equal to one-eighth of China's total carbon dioxide emissions at the time — and contribute significantly to climate change, the authors said. Eighty percent of the projects were set to open in northwestern China, the country's most water-stressed region.

Around the same time, China's National Energy Administration banned the construction of any coal-to-gas plants that would produce less than two billion cubic meters of gas per year.

So while central government officials take a cautious approach to the industry, they are still allowing it to grow. Last year, a coal-to-gas plant opened in Henan Province, the first in the east. Also in 2016, the Ministry of Environmental Protection approved three new plants. And at least three existing plants are expanding.

Kingho signed an agreement last October to push ahead with a "Phase 2" expansion of the Yining plant.

Kingho declined an interview request and did not agree to take me on a tour. On a Saturday, I walked onto the grounds, after a guard in a black uniform at the main gate asked me to sign a visitors' logbook. The roads were wide and quiet, and a few workers in uniforms rode around on electric scooters.

The plant is far from China's largest population centers. Across the northwest, companies would have to build and operate long pipelines to get gas to cities like Beijing and Shanghai. Yet local governments and state-owned enterprises justify construction of the plants by saying they will eventually supply gas to eastern China.

The idea of transporting gas from the west aligns with pollution control policies announced in 2013. That year, after a rise in public anger over air pollution, the government said officials in populous parts of eastern China had to limit coal use, which meant those officials would have to look for other energy sources, including gas.

State-owned energy companies label coal-to-gas plants as "clean energy" or "new energy," despite their high levels of carbon dioxide emissions. In his 2015 article, Mr. Li said that carbon dioxide emissions from coal-to-gas were 270 percent greater than natural gas from traditional sources.

"They do not reduce emissions," said Lin Boqiang, the director of the China Center for Energy Economics Research at Xiamen University. "They only shift emissions elsewhere. Actually, they increase emissions."

Another climate researcher, Jiang Kejun, said that if China adopted coal-to-gas on a larger scale in five to 10 years, then officials would have to find a way to get companies to use carbon capture technology to limit the effect of the emissions.

Since coal-to-gas plants are expensive to operate, market forces may end up dampening their popularity, he said.

"It's very risky, and there is no clear future," he said. "So I believe enterprises themselves will be very cautious."

In early 2015, China Business Journal published a report that said a coal-to-gas plant run by a Huineng Group subsidiary in Ordos, Inner Mongolia, was operating at a daily loss of 1.12 million renminbi, about $160,000.

Antipollution regulations, if properly enforced by officials, could also bring the projects to a halt. The Yili Xintian Coal Chemical Company has been trying to open a new coal-to-gas plant here in Yining, but local environmental protection officials told the company to stop in 2015. They said the plant could pollute the Yili River, an important water source.

In February 2016, the company rewrote the project report, arguing that the plant would be environmentally friendly. Despite objections from local residents during the public comment period, the Ministry of Environmental Protection approved the plant in December.

China has forged recent agreements with Russia and Central Asian nations to buy natural gas. Pipelines running across the Asian heartland transport the gas through Xinjiang to eastern China.

Those pipelines also damage the environment and cultural sites, but some scholars say they have greater concerns over the use of coal-to-gas.

"The development of coal-to-gas technology," Mr. Li wrote, "will actually bring a relatively large negative influence to the already fragile local ecological environment."

VANESSA PIAO and **KAROLINE KAN** contributed research from Beijing.

Tropical Forests Suffered Near-Record Tree Losses in 2017

BY BRAD PLUMER | JUNE 27, 2018

IN BRAZIL, FOREST FIRES set by farmers and ranchers to clear land for agriculture raged out of control last year, wiping out more than 3 million acres of trees as a severe drought gripped the region. Those losses undermined Brazil's recent efforts to protect its rain forests.

In Colombia, a landmark peace deal between the government and the country's largest rebel group paved the way for a rush of mining, logging and farming that caused deforestation in the nation's Amazon region to spike last year.

And in the Caribbean, Hurricanes Irma and Maria flattened nearly one-third of the forests in Dominica and a wide swath of trees in Puerto Rico last summer.

In all, the world's tropical forests lost roughly 39 million acres of trees last year, an area roughly the size of Bangladesh, according to a report Wednesday by Global Forest Watch that used new satellite data from the University of Maryland. Global Forest Watch is part of the World Resources Institute, an environmental group.

That made 2017 the second-worst year for tropical tree cover loss in the satellite record, just below the losses in 2016.

The data provides only a partial picture of forest health around the world, since it does not capture trees that are growing back after storms, fires or logging. But separate studies have confirmed that tropical forests are shrinking overall, with losses outweighing the gains.

The new report comes as ministers from forest nations around the world meet in Oslo this week to discuss how to step up efforts to protect the world's tropical forests, which host roughly half of all species worldwide and play a key role in regulating Earth's climate.

"These new numbers show an alarming situation for the world's rain forests," said Andreas Dahl-Jorgensen, deputy director of the

Norwegian government's International Climate and Forest Initiative. "We simply won't meet the climate targets that we agreed to in Paris without a drastic reduction in tropical deforestation and restoration of forests around the world."

TRACKING FOREST LOSS

Trees, particularly those in the lush tropics, pull carbon dioxide out of the air as they grow and lock that carbon in their wood and soil. When humans cut down or burn trees, the carbon gets released back into the atmosphere, warming the planet. By some estimates, deforestation accounts for more than 10 percent of humanity's carbon dioxide emissions each year.

But figuring out precisely where forests are vanishing has long been a challenge. For decades, the United Nations' Food and Agriculture Organization has relied on ground-level assessments from individual countries to track deforestation. Yet not all tropical countries have adequate capacity to monitor their forests, and the measurements can be plagued by inconsistencies.

In 2013, scientists at the University of Maryland unveiled a fresh approach. Using satellite data recently made free, they have been tracking changes in tree canopy area around the world. This method has its own limits: More work is still needed to distinguish between trees that are being intentionally harvested in plantations and those that are being newly cleared in older, natural forests. The latter is a much bigger concern for habitat loss and climate change.

Both ground-level assessments and satellite data are important, said Matthew C. Hansen, a scientist who leads the monitoring effort at the University of Maryland. "But what satellites can do is identify disturbances much more quickly," he said. "We can map the first logging road into a forest and then send out an alert."

CONCERN IN COLOMBIA, BRAZIL AND CONGO

From the satellite imagery, researchers noticed that Colombia lost 1 million acres of forest in 2017, a stunning 46 percent uptick from the previous year. Many of these losses took place in the Colombian Amazon, in areas that used to be strictly controlled by the Revolutionary Armed Forces of Colombia, or FARC, a guerrilla group that imposed tight controls on logging and land-clearing but disarmed last year amid a landmark peace accord.

"As FARC has demobilized, large areas are opening up once again, and you're seeing this rush of people grabbing land for different reasons, like planting cocoa or cattle ranching," said Mikaela Weisse, a research analyst with Global Forest Watch.

She added that the Colombian government recently announced new policies to work with indigenous communities to protect forests, but said it was too early to declare success.

The satellite data also provided a clearer picture of Brazil's vast Amazon rain forests, long vulnerable to widespread deforestation. Over the past decade, the Brazilian government has moved to reduce illegal logging, and Western agriculture companies like Cargill have pledged to farm more sustainably.

But the Global Forest Watch analysis showed that Brazil lost a record amount of tree cover in 2016 and 2017, in part because of large fire outbreaks in the Amazon. These fires are typically started by farmers and ranchers to clear land, but a severe drought last year caused them to spread rapidly, particularly in the parched southeast. The satellites also picked up evidence of large-scale land-clearing that may be occurring in areas where enforcement is weak.

"The big concern is that we're starting to see a new normal, where fires, deforestation, drought and climate change are all interacting to make the Amazon more flammable," Ms. Weisse said.

Elsewhere in the world, the satellite data showed that the Democratic Republic of Congo last year saw more forest loss than any other country outside of Brazil — some 3.6 million acres, up 6 percent from the previous year — with small-scale logging, charcoal production and farming all likely playing key roles.

POSSIBLE PROGRESS IN INDONESIA

The researchers did find a tentative bright spot in Indonesia, where a government crackdown on deforestation may be showing early signs of success.

Over the past several decades, Indonesia's farmers have been draining and burning the country's peatlands — thick layers of partially decomposed vegetation that hold enormous stores of carbon — in order to grow crops like palm oil. But in 2015, amid a strong El Niño and severe dry spell, the country had its worst fire season in decades, blanketing Southeast Asia in deadly smoke.

In 2016, Indonesia's government imposed a new moratorium on the conversion of peatland, while Norway pledged $50 million for enforcement. Early signs are encouraging: primary forest loss on Indonesia's protected peatland dropped 88 percent in 2017, to the lowest level in years. Still, experts said, the real test of success may come when the next El Niño hits.

But such positive stories tend to be a relative rarity and experts say much more is needed to slow the pace of deforestation. To date, just 2 percent of international financing for activities to fight climate change goes toward forest conservation, said Frances Seymour, a senior fellow at the World Resources Institute.

"We're trying to put out a house fire with a teaspoon," she said.

BRAD PLUMER is a reporter covering climate change, energy policy and other environmental issues for The Times's climate team.

The Amazon on the Brink

OPINION | BY PHILIP FEARNSIDE AND RICHARD SCHIFFMAN | SEPT. 26, 2018

Once a leader in protecting the region's vast forests, Brazil is now moving in the opposite direction.

THE TRUMP ADMINISTRATION is not the only government that has been busy slashing funds for environmental protection. Brazil has been doing the same.

While Mr. Trump makes no bones about his desire to roll back environmental laws, Brazil's president, Michel Temer, a signatory of the Paris climate agreement, has sent mixed signals. To his credit, Mr. Temer pledged in Paris to cut his country's carbon dioxide emissions 37 percent below 2005 levels by 2025.

His actions since then tell a different story. Last year, the Environment Ministry's budget was cut nearly in half, as part of a national austerity plan amid Brazil's punishing recession. And the agency responsible for protecting Brazil's vast system of indigenous reserves is being virtually dismantled by draconian staff cuts.

Funding for critical law enforcement to protect the rain forest from illegal timber cutting has also been decimated. In 2017, Brazil was the most dangerous country in the world for people defending the land or the environment, according to a tally by the group Global Witness, in collaboration with The Guardian newspaper. Forty-six people died. (The Pastoral Land Commission, a private advocacy group in Brazil for the rural poor, said at least 65 rural activists were murdered in disputes over development.)

If the government's retrenchment on environmental protection continues, there may soon be nothing to stop the chain saws on the Amazonian frontier, where the rule of law can be weak and land is frequently seized and cleared illegally. This has implications beyond Brazil. The Amazon's lush forests make up the largest reserve of

LALO DE ALMEIDA FOR THE NEW YORK TIMES

Deforestation in Maranhão State in northeastern Brazil. Brazil lost 2,682 square miles of Amazonian forests in 2017.

carbon dioxide on the surface of the earth. This potent greenhouse gas is released when forests are burned or bulldozed and left to decay.

Brazilians are going to the polls on Oct. 7 to choose among nine candidates for the presidency, with a likely runoff later in the month between the top two vote-getters. The outcome will bear significantly on the future of the Amazon.

The current front-runner, Jair Bolsonaro, is a climate-change skeptic who has been called "the tropical Trump." He has threatened to take Brazil out of the Paris climate accord. Another of the leading contenders, the former São Paulo mayor Fernando Haddad, is regarded as a moderate on the environment. Marina Silva, who as Brazil's environment minister pushed to limit deforestation and encourage sustainable development in the Amazon, is running well behind in the latest polls.

LALO DE ALMEIDA FOR THE NEW YORK TIMES

A chain saw used for illegal logging in the Alto Rio Guamá Indigenous Territory in Brazil, abandoned by loggers who fled when government agents raided the site last year.

Deforestation rates have been trending mostly upward since 2012 and will surely escalate if a raft of proposed laws and regulatory changes to weaken environmental protections are enacted. Brazil lost 2,682 square miles of Amazonian forests in 2017. That is almost nine times the size of New York City and 78 percent above the government's own target for meetings its obligation under the Paris accord.

In an analysis published in July in the journal Nature Climate Change, 10 Brazilian scientists concluded that "the abandonment of deforestation control policies and the political support for predatory agricultural practices" will make it impossible for Brazil to reduce carbon dioxide emissions to the level the country promised in Paris. Continued weak environmental governance, the scientists warned, could lead to the loss of up to 17,000 square miles of rain forest a year, endangering the entire Amazon ecosystem.

Why this change in policy? The scientists put it succinctly: "In exchange for political support, the Brazilian government is signaling landholders to increase deforestation."

President Temer's minister of justice is pushing plans to allow agribusiness to rent indigenous land that had been off limits to developers. Other proposals would effectively freeze the creation of new protected areas, open others to resource exploitation and block the mapping of boundaries of indigenous lands, potentially opening native communities and their forests to invasion by miners and ranchers.

Indigenous territories contain more forest than all of the government's conservation units combined, and historically Brazil's native peoples have been far more effective in defending the rain forest than the government or private landowners.

The anti-environment agenda is being pushed by a coalition of large landowners and agribusinesses in Congress (the "bancada ruralista" or so-called ruralists). Regular revelations of corruption involving government ministers, legislators — and also, President Temer himself — have provided them with cover to pursue regressive measures, like a proposed constitutional amendment that would prevent Brazil's regulators from blocking environmentally unsound road and development projects.

Scores of such projects planned for inaccessible regions of the Amazon may be fast-tracked should the amendment pass and the environmental review process is gutted as a result, as now seems likely. For example, the 540-mile long BR-319 highway would, if completed, open a vast area in the central and northern Amazonia to deforestation.

Not only is little being done to prevent illegal land use, some laws have encouraged it. Last year, the "grileiro," or land-grabber, law legalized tracts of nearly 6,200 acres that were taken illegally — a boon to land speculators and others who seize public lands for their own use.

Not so long ago, Brazil was doing things right. Despite low global prices for soy and beef, the nation experienced a remarkable economic expansion while deforestation fell 60 percent from 2004 to 2007, demonstrating that environmental growth is consistent with economic growth. But now that demand for soy and beef on the global market is high, pressure on the forest is mounting. Deforestation is still well below historical highs. But that could soon change if the power of the agribusiness lobby is not checked.

The ruralists came into increased prominence toward the end of the administration of the Workers Party president, Luiz Inácio Lula da Silva, and grew more powerful during the term of his successor, Dilma Rousseff. During Ms. Rousseff's presidency, which ended when she was removed from office following impeachment, amnesty was granted to landowners who had illegally cleared forests, encouraging continued lawbreaking in the world's largest rain forest.

Climate change has also increased the danger of catastrophic forest fires. In dry El Niño years, burned areas can greatly exceed what is cleared for cattle pasture.

The changing climate represents both a threat to the Amazon and a key reason for protecting its forests. Transpiration from tree leaves generates rivers of moisture in the atmosphere that act as conveyor belts bringing much-needed rain to Brazil's heavily populated south and to Argentina. São Paulo, which already regularly runs dangerously short of water, will be a major victim if continued deforestation removes this water vapor transport that fills the reservoirs on which the city, South America's largest, depends.

The Amazon forest itself would also suffer. As more of it is cut, the huge volume of self-generated rainfall it needs to remain verdant is steadily reduced, worsening droughts made more frequent and severe by global warming. At some point — we are not sure how close we are to this critical tipping point — the entire ecosystem dries out.

Brazilians have consistently said in opinion polls that they want to preserve the Amazon. But in the current atmosphere of unbridled greed and corruption in high places, the voices for wise policy are often drowned out.

Brazil alone has not created the deforestation problem, and neither can it address it alone. Demand for Brazil's beef from Western nations, and increasingly from China, has created an enormous temptation to cut the forest to turn a quick profit.

Importing nations and Brazilian soy traders and beef producers must live up to their pledges that they will not buy products produced on cleared forest. And global financial institutions must stop funding projects that result in deforestation. They should also increase their assistance to Brazil and other tropical countries to help them maintain their forests and pursue nondestructive alternatives to cutting them.

Only then, can we insure that the Amazonia forests, the living heart of Brazil — and of the world — will remain intact.

PHILIP FEARNSIDE is an ecologist at the National Institute for Research in Amazonia in Brazil. **RICHARD SCHIFFMAN** is an environmental journalist.

CHAPTER 4

Climate Skepticism Becomes a Partisan Opportunity

Because climate change works gradually and seemingly at random, not everyone has accepted the scientific consensus. And as the effects of climate change have become visible, skepticism and denial have only become more influential in the public debate. Prominent politicians like James Inhofe and Scott Pruitt, often with financial ties to fossil fuel industries, have turned climate denial into a partisan issue. While this denial has helped block action on the problem, climate skepticism raises even deeper questions about the role of doubt in scientific inquiry.

Material Shows Weakening of Climate Reports

BY ANDREW C. REVKIN AND MATTHEW L. WALD | MARCH 20, 2007

WASHINGTON, MARCH 19 — A House committee released documents Monday that showed hundreds of instances in which a White House official who was previously an oil industry lobbyist edited government climate reports to play up uncertainty of a human role in global warming or play down evidence of such a role.

In a hearing of the House Committee on Oversight and Government Reform, the official, Philip A. Cooney, who left government in 2005, defended the changes he had made in government reports over several

choice of words by his colleague was poor but noted that scientists often used the word "trick" to refer to a good way to solve a problem, "and not something secret."

At issue were sets of data, both employed in two studies. One data set showed long-term temperature effects on tree rings; the other, thermometer readings for the past 100 years.

Through the last century, tree rings and thermometers show a consistent rise in temperature until 1960, when some tree rings, for unknown reasons, no longer show that rise, while the thermometers continue to do so until the present.

Dr. Mann explained that the reliability of the tree-ring data was called into question, so they were no longer used to track temperature fluctuations. But he said dropping the use of the tree rings was never something that was hidden, and had been in the scientific literature for more than a decade. "It sounds incriminating, but when you look at what you're talking about, there's nothing there," Dr. Mann said.

In addition, other independent but indirect measurements of temperature fluctuations in the studies broadly agreed with the thermometer data showing rising temperatures.

Dr. Jones, writing in an e-mail message, declined to be interviewed.

Stephen McIntyre, a blogger who on his Web site, climateaudit.org, has for years been challenging data used to chart climate patterns, and who came in for heated criticism in some e-mail messages, called the revelations "quite breathtaking."

But several scientists whose names appear in the e-mail messages said they merely revealed that scientists were human, and did nothing to undercut the body of research on global warming. "Science doesn't work because we're all nice," said Gavin A. Schmidt, a climatologist at NASA whose e-mail exchanges with colleagues over a variety of climate studies were in the cache. "Newton may have been an ass, but the theory of gravity still works."

He said the breach at the University of East Anglia was discovered after hackers who had gained access to the correspondence sought

Tuesday to hack into a different server supporting realclimate.org, a blog unrelated to NASA that he runs with several other scientists pressing the case that global warming is true.

The intruders sought to create a mock blog post there and to upload the full batch of files from Britain. That effort was thwarted, Dr. Schmidt said, and scientists immediately notified colleagues at the University of East Anglia's Climatic Research Unit. The first posts that revealed details from the files appeared Thursday at The Air Vent, a Web site devoted to skeptics' arguments.

At first, said Dr. Michaels, the climatologist who has faulted some of the science of the global warming consensus, his instinct was to ignore the correspondence as "just the way scientists talk."

But on Friday, he said that after reading more deeply, he felt that some exchanges reflected an effort to block the release of data for independent review.

He said some messages mused about discrediting him by challenging the veracity of his doctoral dissertation at the University of Wisconsin by claiming he knew his research was wrong. "This shows these are people willing to bend rules and go after other people's reputations in very serious ways," he said.

Spencer R. Weart, a physicist and historian who is charting the course of research on global warming, said the hacked material would serve as "great material for historians."

Energy Firms in Secretive Alliance With Attorneys General

BY ERIC LIPTON | DEC. 6, 2014

THE LETTER TO the Environmental Protection Agency from Attorney General Scott Pruitt of Oklahoma carried a blunt accusation: Federal regulators were grossly overestimating the amount of air pollution caused by energy companies drilling new natural gas wells in his state.

But Mr. Pruitt left out one critical point. The three-page letter was written by lawyers for Devon Energy, one of Oklahoma's biggest oil and gas companies, and was delivered to him by Devon's chief of lobbying.

"Outstanding!" William F. Whitsitt, who at the time directed government relations at the company, said in a note to Mr. Pruitt's office. The attorney general's staff had taken Devon's draft, copied it onto state government stationery with only a few word changes, and sent it to Washington with the attorney general's signature. "The timing of the letter is great, given our meeting this Friday with both E.P.A. and the White House."

Mr. Whitsitt then added, "Please pass along Devon's thanks to Attorney General Pruitt."

The email exchange from October 2011, obtained through an open-records request, offers a hint of the unprecedented, secretive alliance that Mr. Pruitt and other Republican attorneys general have formed with some of the nation's top energy producers to push back against the Obama regulatory agenda, an investigation by The New York Times has found.

Attorneys general in at least a dozen states are working with energy companies and other corporate interests, which in turn are providing them with record amounts of money for their political campaigns, including at least $16 million this year.

They share a common philosophy about the reach of the federal government, but the companies also have billions of dollars at stake.

DYLAN HOLLINGSWORTH FOR THE NEW YORK TIMES

The Oklahoma attorney general, second from right, in Dallas in July, and his Republican counterparts have formed alliances to oppose federal regulations.

And the collaboration is likely to grow: For the first time in modern American history, Republicans in January will control a majority — 27 — of attorneys general's offices.

The Times reported previously how individual attorneys general have shut down investigations, changed policies or agreed to more corporate-friendly settlement terms after intervention by lobbyists and lawyers, many of whom are also campaign benefactors.

But the attorneys general are also working collectively. Democrats for more than a decade have teamed up with environmental groups such as the Sierra Club to use the court system to impose stricter regulation. But never before have attorneys general joined on this scale with corporate interests to challenge Washington and file lawsuits in federal court.

Out of public view, corporate representatives and attorneys general are coordinating legal strategy and other efforts to fight federal regulations, according to a review of thousands of emails and court documents and dozens of interviews.

"When you use a public office, pretty shamelessly, to vouch for a private party with substantial financial interest without the disclosure of the true authorship, that is a dangerous practice," said David B. Frohnmayer, a Republican who served a decade as attorney general in Oregon. "The puppeteer behind the stage is pulling strings, and you can't see. I don't like that. And when it is exposed, it makes you feel used."

For Mr. Pruitt, the benefits have been clear. Lobbyists and company officials have been notably solicitous, helping him raise his profile as president for two years of the Republican Attorneys General Association, a post he used to help start what he and allies called the Rule of Law campaign, which was intended to push back against Washington.

That campaign, in which attorneys general band together to operate like a large national law firm, has been used to back lawsuits and other challenges against the Obama administration on environmental issues, the Affordable Care Act and securities regulation. The most recent target is the president's executive action on immigration.

"We are living in the midst of a constitutional crisis," Mr. Pruitt told energy industry lobbyists and conservative state legislators at a conference in Dallas in July, after being welcomed with a standing ovation. "The trajectory of our nation is at risk and at stake as we respond to what is going on."

Mr. Pruitt has responded aggressively, and with a lot of helping hands. Energy industry lobbyists drafted letters for him to send to the E.P.A., the Interior Department, the Office of Management and Budget and even President Obama, The Times found.

Industries that he regulates have also joined him as plaintiffs in court challenges, a departure from the usual role of the state attorney general, who traditionally sues companies to force compliance with state law.

Energy industry lobbyists have also distributed draft legislation to attorneys general and asked them to help push it through state legislatures to give the attorneys general clearer authority to challenge the Obama regulatory agenda, the documents show.

"It is quite new," said Paul Nolette, a political-science professor at Marquette University and the author of the forthcoming book "Federalism on Trial: State Attorneys General and National Policy Making in Contemporary America." "The scope, size and tenor of these collaborations is, without question, unprecedented."

And it is an emerging practice that several former attorneys general say threatens the integrity of the office.

"It is a magnificent and noble institution, the office of attorney general, as it is truly the lawyer for the people," said Terry Goddard, a Democrat who served two terms as Arizona's attorney general and who, like Mr. Frohnmayer, reviewed copies of the documents collected by The Times. "That independence is clearly at risk here. What is happening diminishes the reputation of individual attorneys general and the community as a group."

Mr. Pruitt, who has emerged as a hero to conservative activists, dismissed this criticism as misinformed.

"Those kinds of questions arise from the environment we are in — a very dysfunctional, distrustful political environment," Mr. Pruitt said in an interview. "I can say to you that is not who we are or have ever been, and despite those criticisms we sit around and make decisions about what is right, and what represents adherence to the rule of law, and we seek to advance that and try to do the best we can to educate people about our viewpoint."

In a state dominated by the energy industry, Mr. Pruitt's stands have been widely popular. "Attorney General Pruitt has been a champion for our state," said State Senator Mike Schulz, a Republican who is the majority floor leader. "The State of Oklahoma is in a better position than the E.P.A. to regulate drilling."

But Mr. Pruitt's ties with industry are clear. One of his closest partners has been Harold G. Hamm, the billionaire chief executive of Continental Resources, which is among the biggest oil and gas drilling companies in both Oklahoma and North Dakota.

This year, Mr. Pruitt joined with a group aligned with Mr. Hamm to sue the Interior Department over its plan to consider adding animals such as the lesser prairie chicken to the endangered species list, a move that Mr. Hamm has said could knock out "some of the most promising land for oil and gas leases in the country." The suit was filed after Mr. Hamm announced that he would serve as the chairman of Mr. Pruitt's re-election campaign.

"Time and time again, General Pruitt has stood up and bravely fought for the rights of Oklahomans in those instances when the federal government has overextended its hand," Mr. Hamm said as his role in Mr. Pruitt's re-election effort was announced.

A POTENT ALLY

Energy industry executives and lobbyists from across the United States saw great potential in Mr. Pruitt, a gifted politician who had been a state legislator and a minor-league baseball team co-owner and executive before running for attorney general.

Among them was Andrew P. Miller, a patrician 81-year-old former Virginia attorney general. Mr. Miller is a regular at gatherings of state attorneys general at resort destinations, and his client list includes TransCanada, the backer of the Keystone XL pipeline; the Southern Company, the Georgia-based electric utility, which has a large number of coal-burning power plants; and the investor group behind the proposed Pebble Mine in Alaska.

For the energy industry, Mr. Pruitt was an easy choice.

"There's a mentality emanating from Washington today that says, 'We know best,' " Mr. Pruitt said during his 2010 campaign. "It's a one-size-fits-all strategy, a command-and-control kind of approach, and we've got to make sure we know how to respond to that."

Among Mr. Pruitt's first acts was to create a "federalism office," which challenged the Obama administration's plan to reduce haze in southwestern Oklahoma by requiring coal-burning electricity plants in the state to install new pollution control equipment.

His interaction with the industry, Mr. Pruitt said during an interview at his Oklahoma City office, has been motivated by a desire to gather information from experts, while defending his state's longstanding tradition of self-determination.

That ethos, he said, is depicted in a large oil painting in his office that shows local authorities with rifles at the ready confronting outsiders during the land rush era. "The founders recognized that power concentrated in a few is a bad thing," Mr. Pruitt said.

Mr. Miller made it his job to promote Mr. Pruitt nationally, both as a spokesman for the Rule of Law campaign and as the president of the Republican Attorneys General Association.

"I regard the general as the A.G. best suited to take this lead on this question of federalism," Mr. Miller wrote to Mr. Pruitt's chief of staff in April 2012. "The touchstone of this initiative would be to organize the states to resist federal 'overreach' whenever it occurs."

To Mr. Miller, having Mr. Pruitt as an advocate fit a broader strategy. He wanted state attorneys general to band together the way they did when they challenged the health care law in 2010. In that effort, they hired a major national corporate law firm, Baker Hostetler, to argue the case, with much of the bill being paid through donations from executives at corporations that oppose the law.

In his initial appeal to Mr. Pruitt, Mr. Miller insisted that his approach was not "client driven." But he soon began to name individual clients — TransCanada and Pebble Mine in Alaska — that he wanted to include in the effort. The E.P.A. has held up the Pebble Mine project, which could potentially yield 80 billion pounds of copper, after concluding it would "threaten one of the world's most productive salmon fisheries."

"This strike force ought to take the form of a national state litigation team to challenge the E.P.A.'s overreach," Mr. Miller said in an email to Mr. Pruitt's office. "Like the Dalmatian at the proverbial firehouse, it could move out smartly when the alarm sounded."

A CALL TO ARMS

Mr. Miller's pitch to Mr. Pruitt became a reality early last year at the historic Skirvin Hilton Hotel in Oklahoma City, where he brought together an extraordinary assembly of energy industry power brokers and attorneys general from nine states for what he called the Summit on Federalism and the Future of Fossil Fuels.

The meeting took place in the shadow of office towers that dominate Oklahoma City's skyline and are home to Continental Resources, a leader in the nation's fastest-growing oil field, the Bakken formation of North Dakota, as well as Devon Energy, which drilled 1,275 new wells last year.

More liberal attorneys general, such as Douglas F. Gansler, Democrat of Maryland, did not participate.

"Indeed, General Gansler would in all likelihood try to hijack your summit," Mr. Miller wrote to Mr. Pruitt in an email. "At best you would be left to preside over a debate, rather than a call to arms."

Oklahoma energy companies were there, according to an agenda, joined by executives from Peabody Energy of Missouri, the world's largest private-sector coal producer, as well as the Southern Company, which has aggressively challenged federal air pollution mandates.

The nation's top corporate energy regulatory lawyers were there, too, including F. William Brownell, a senior partner at the law firm Hunton & Williams, which has spent more than 25 years fighting the enforcement of the Clean Air Act.

The event was organized by an energy-industry-funded law and economics center at George Mason University of Virginia. The center is part of the brain trust of conservative, pro-industry groups that have worked from the sidelines to help Mr. Pruitt and other attorneys general.

And there was nothing ambiguous about the agenda.

"Suggested Responses to Assaults on Federalism" was the topic of one breakfast meeting, moderated by Attorney General Wayne K.

Stenehjem of North Dakota, that showcased Mr. Brownell and three other top corporate regulatory lawyers. Mr. Hamm was the featured dinner speaker.

"We need to ensure the robust role of the states," said Paul M. Seby, another coal industry lawyer who attended. "And as the chief law enforcement officers, it is not surprising this is becoming a cornerstone of attorney generals' attention."

Attorneys general said they had no choice but to team up with corporate America. "When the federal government oversteps its legal authority and takes actions that hurt our businesses and residents, it's entirely appropriate for us to partner with the adversely affected private entities in fighting back," said Attorney General Pam Bondi of Florida, whose top deputy attended the meeting.

A 'STRIKE FORCE'

The impact of the gathering was immediate. A week later, a new Federalism in Environmental Policy task force was established by lawyers in the offices of 19 state attorneys general, according to email records obtained from the office of Attorney General Timothy C. Fox of Montana, who had participated in the Oklahoma meeting.

"This message is in follow-up to the excellent environmental conference put on last week by George Mason University and hosted by the Oklahoma attorney general's office," said one email sent by Katie Spohn, the deputy attorney general in Nebraska. "In order to continue our coordination of efforts regarding Federalism in Environmental Policy, I am seeking input from each state who participated in the conference."

Mr. Miller was pleased. "Just the kind of strike force I was talking about," he said in an interview.

And the input poured forth. The states worked to detail major federal environmental action, like efforts to curb fish kills, reduce ozone pollution, slow climate change and tighten regulation of coal ash. Then they identified which attorney general's office was best positioned to try to monitor it and, if necessary, attempt to block it.

Follow-up by Mr. Pruitt's federalism office often came after coordination with industry representatives, especially from Devon Energy. The company, one of the most important financial supporters for the Republican Attorneys General Association, is guarded about its public profile. But it readily turned to Mr. Pruitt and his staff for help, setting up meetings for the attorney general with its chief executive, its chief lobbyist and other important players.

"We have a clear obligation to our shareholders and others to be involved in these discussions," John Porretto, a Devon spokesman, said in a statement.

While some of the exchanges were general in character, others were quite explicit, especially the communication about the E.P.A.'s methane regulations that had prompted Mr. Whitsitt, the Devon official, to propose that Mr. Pruitt send a letter to the agency. "Just a note to pass along the electronic version of the draft letter to Lisa Jackson at E.P.A.," said one September 2011 letter to Mr. Pruitt's chief of staff from Mr. Whitsitt. "We have no pride of authorship, so whatever you do on this is fine."

Mr. Pruitt took the letter and, after changing just 37 words in the 1,016-word draft, copied it onto his state government letterhead and sent it to Ms. Jackson, the E.P.A. administrator.

That was just one of his challenges to Washington. Devon officials also turned to Mr. Pruitt to enlist other Republican attorneys general and Republican governors to oppose a rule proposed by the Bureau of Land Management that would regulate hydraulic fracturing, or fracking, on federal land.

"As promised, we are sending you the attached draft of the R.G.A./RAGA follow-up letter to President Obama opposing B.L.M.'s proposed rule," Brent Rockwood, Devon's director of government affairs, wrote to Mr. Pruitt's staff in late 2012, in an email marked "confidential."

Weeks later, that letter was sent to Mr. Obama with only a few word changes, signed by Mr. Pruitt and Gov. Bobby Jindal of Louisiana, who was the head of the Republican Governors Association at the time.

Company officials again expressed their pleasure to Mr. Pruitt.

"I've learned that we're having an effect — and may be able to have more, perhaps even to having the rule withdrawn or shifted to almost a reporting-only one," Mr. Whitsitt wrote, in another email marked "confidential."

The rule — which the industry claims would cost $346 million a year to comply with — has still not been issued.

Coordination between the corporations and teams of attorneys general involved in the Rule of Law effort also involves actual litigation to try to clear roadblocks to energy projects, documents show. Energy producers, for instance, wanted to sue the Interior Department as it considered adding animals such as the sage grouse — which nests near sites of oil and gas drilling — to a list of endangered species, a move that could put tens of thousands of acres off limits to new drilling.

The energy companies could have sued on their own, but their executives believed that the case would be more potent by bringing in Mr. Pruitt and the weight of the State of Oklahoma.

"We just came to the conclusion he would be the best person to be the lead attorney on this," said Mike McDonald, an owner of Triad Energy, a small oil and gas exploration company, and the president of a group that calls itself the Domestic Energy Producers Alliance. "He has exceeded our expectations."

For the industry, the state is an extremely valued partner because states are granted "special solicitude" from the federal courts, a critical advantage to private companies that helps confer legal standing and means that a matter is less likely to be dismissed.

Mr. Pruitt's office, in a statement to The Times, rejected any suggestion that the attorney general has been wrong to send to Washington comment letters written by industry lobbyists, or to take up their side in litigation.

"The A.G.'s office seeks input from the energy industry to determine real-life harm stemming from proposed federal regulations or actions," the statement said. "It is the content of the request not the source of the request that is relevant."

Persuading lawmakers to offer legislation has been another effective lobbying tool. In West Virginia, Mr. Miller handed Attorney General Patrick Morrisey a draft of legislation that he argued would put West Virginia in a better position to sue the Obama administration over proposed regulations to tighten pollution controls on power plants, emails show.

"I trust you will find the legislation acceptable in its present form," Mr. Miller wrote to Mr. Morrisey in February, referring to a private meeting the two had had in the law library of Mr. Morrisey's office in Charleston. "If so, I would appreciate your having it introduced by your friends in both the Senate and the House."

A version of the bill was introduced and passed by the West Virginia Legislature in March. Delegate Rupert Phillips Jr., the chief sponsor of a second bill that also contained language identical to what Mr. Miller had requested, said in an interview that he had acted with Mr. Morrisey's support, an account supported by William B. Raney, the president of the West Virginia Coal Association.

"It is nice to have everybody singing from the same sheet of music," Mr. Raney said.

A spokesman for Mr. Morrisey disputed this account, saying that while he supported the effort to challenge the rule, he did not play a role in promoting the legislation.

BLURRED LINES

The work in Mr. Pruitt's office has sometimes seemed to blur the distinction between his official duties and the advancement of his political career.

Mr. Pruitt's chief of staff, Crystal Drwenski, served as gatekeeper to his office, arranging meetings and helping companies get Mr. Pruitt and his staff to intervene with the federal authorities. But Ms. Drwenski also played an important supplemental role for the attorney general: fund-raising aide.

"A.G. Pruitt is working with the Republican Attorneys General Association on their national meeting in Washington," Ms. Drwenski wrote to Mr. Whitsitt. "The benefit of membership and participation is having 25 Republican A.G.s in a room to discuss policy issues."

Ms. Drwenski wanted Devon Energy's help in enlisting the American Petroleum Institute, and Mr. Whitsitt agreed.

"I've put in a plug to A.P.I.," Mr. Whitsitt wrote back to Ms. Drwenski, a few hours after her request, having reached out to the organization's senior lobbyist, Marty Durbin. "He is expecting a call."

In addition to the American Petroleum Institute, major energy companies — ConocoPhillips, the oil and gas company; Alpha Natural Resources, a coal mining giant; and American Electric Power, the nation's biggest coal consumer — have recently joined the Republican Attorneys General Association, bringing in hundreds of thousands of additional dollars to the group, internal documents show.

By last year, the association was starting to pull in so much money under Mr. Pruitt's leadership that it decided to break free from its partnership with the Republican State Leadership Committee, a group that represents state elected officials. Within months, the association also set up the Rule of Law Defense Fund, yet another legal entity that allows companies benefiting from the actions of Mr. Pruitt and other Republican attorneys general to make anonymous donations, in unlimited amounts. Fund-raising skyrocketed.

The $16 million that the association has collected this year is nearly four times the amount it collected in 2010, money it used mostly to buy millions of dollars' worth of television advertisements in states like Arizona, Arkansas, Colorado and Nevada, all places where Republican candidates for attorney general won election.

The fund-raising has taken place on the state level as well. Oklahoma Gas & Electric — a for-profit utility that Mr. Pruitt joined with in federal court to fight the E.P.A. — invited its employees to the Petroleum Club in downtown Oklahoma City late last year for a fund-raising event for Mr. Pruitt, drawing donations from about 45 company employees, including the chief executive. Four days later, Mr. Pruitt filed a new appeal in the case — timing that the utility said was a coincidence.

While Mr. Pruitt's efforts to raise money for the Republican Attorneys General Association have been an unqualified success, the lawsuits and regulatory appeals he has filed have yielded mixed results.

In May, the Supreme Court declined to take up the appeal on the Oklahoma Gas & Electric matter, meaning the company is now moving ahead on retrofitting its coal-burning plants. But other lawsuits are pending, including Mr. Pruitt's challenge of the Dodd-Frank law, which rewrote the nation's financial regulations, and, perhaps most important, his challenge of the tax subsidies that are a critical part of the Obama administration's health care law.

Mr. Pruitt's staff has juggled various duties — helping major corporations push their challenges against Washington, and then turning to these same executives, at times, to ask them for financial support.

For example, Ms. Drwenski, who is no longer Mr. Pruitt's chief of staff, asked Devon Energy in 2012, on a workday afternoon, for help in signing up the American Petroleum Institute as a member of the Republican Attorneys General Association.

She used her personal email account to send out the initial request. But the subsequent exchange took place on her work email account, even though Oklahoma state law prohibits state officials from using state property or time to solicit political contributions. A spokesman for Mr. Pruitt said, "It is entirely possible she could have been taking a late lunch."

Mr. Pruitt, who ran unopposed to win a second term, has not needed much of the money himself, but his fund-raising efforts have greatly benefited other Republicans running for the job.

That explains the partylike atmosphere late last month in South Florida, where members of the Republican Attorneys General Association held their fall meeting at the chic Fontainebleau Miami Beach, along with hundreds of lobbyists, lawyers and corporate executives, whose companies had paid as much as $125,000 for the privilege to celebrate with them.

NICK OXFORD FOR THE NEW YORK TIMES

An Oklahoma drilling site.

During the opening reception, on a giant terrace overlooking the Atlantic Ocean, with red, white and blue lights beaming onto the walls and rock music blasting, the Republican attorneys general strode to the stage to trumpet their new majority in the states.

Mr. Pruitt was there for the weekend's festivities, an event at which Devon Energy served as a corporate host, with banners hung in the hotel hallways featuring the corporate logo.

The Oklahoma attorney general's stay was brief. The Rule of Law campaign had a new and urgent target.

"Our president sees himself as above the law," Mr. Pruitt said from Oklahoma City as he announced several days later yet another front in the campaign, a lawsuit he planned to file to challenge the Obama administration's new immigration policies. "We will take action to hold him accountable."

NICK MADIGAN contributed reporting.

E.P.A. Chief Doubts Consensus View of Climate Change

BY CORAL DAVENPORT | MARCH 9, 2017

WASHINGTON — Scott Pruitt, the head of the Environmental Protection Agency, said on Thursday that carbon dioxide was not a primary contributor to global warming, a statement at odds with the established scientific consensus on climate change.

Asked his views on the role of carbon dioxide, the heat-trapping gas produced by burning fossil fuels, in increasing global warming, Mr. Pruitt said on CNBC's "Squawk Box" that "I think that measuring with precision human activity on the climate is something very challenging to do and there's tremendous disagreement about the degree of impact, so, no, I would not agree that it's a primary contributor to the global warming that we see."

"But we don't know that yet," he added. "We need to continue the debate and continue the review and the analysis."

Mr. Pruitt's statement contradicts decades of research and analysis by international scientific institutions and federal agencies, including the E.P.A. His remarks on Thursday, which were more categorical than similar testimony before the Senate, may also put him in conflict with laws and regulations that the E.P.A. is charged with enforcing.

His statements appear to signal that the Trump administration intends not only to roll back President Barack Obama's climate change policies, but also to wage a vigorous attack on their underlying legal and scientific basis.

A report in 2013 by the Intergovernmental Panel on Climate Change, a group of about 2,000 international scientists that reviews and summarizes climate science, found it to be "extremely likely" that more than half the global warming that occurred from 1951 to 2010 was a consequence of human emissions of carbon dioxide and other greenhouse gases.

A January report by NASA and the National Oceanic and Atmospheric Administration concluded, "The planet's average surface temperature has risen about 2.0 degrees Fahrenheit (1.1 degrees Celsius) since the late 19th century, a change driven largely by increased carbon dioxide and other human-made emissions into the atmosphere."

Benjamin D. Santer, a climate researcher at the Energy Department's Lawrence Livermore National Laboratory, said, "Mr. Pruitt has claimed that carbon dioxide caused by human activity is not 'the primary contributor to the global warming that we see.' Mr. Pruitt is wrong."

"The scientific community has studied this issue for decades," Dr. Santer added. "The consensus message from many national and international assessments of the science is pretty simple: Natural factors can't explain the size or patterns of observed warming. A large human influence on global climate is the best explanation for the warming we've measured and monitored."

The basic science showing that carbon dioxide traps heat at the Earth's surface dates to the 19th century, and has been confirmed in many thousands of experiments and observations since.

Mr. Pruitt has faced frequent criticism for his close ties to fossil fuel companies. In his previous job as the attorney general of Oklahoma, he sought to use legal tools to fight environmental regulations on the oil and gas companies that are a major part of the state's economy. A 2014 investigation by The New York Times found that energy lobbyists had drafted letters for Mr. Pruitt to send, on state stationery, to the E.P.A., the Interior Department, the Office of Management and Budget and even Mr. Obama, outlining the economic hardship caused by the environmental rules.

But in a sign of how far outside mainstream views Mr. Pruitt's remarks on Thursday have placed him, even executives of some of the nation's largest fossil fuel producers said they were surprised by his comments. Interviewed at CERAweek, an annual conference of major energy producers this week in Houston, Wael Sawan, an executive vice president at Shell Energy Resources, said he was "absolutely convinced CO2 can cause serious damage to not only this generation but future generations."

Mr. Pruitt spoke at the Houston energy conference on Thursday afternoon in a session moderated by Daniel Yergin, a prominent energy economist and a member of a White House advisory panel. Mr. Yergin did not ask Mr. Pruitt any questions about his remarks from Thursday morning, saying later that he had been unaware of the comments when interviewing him.

Mr. Pruitt did not clarify his comments or respond to reporters who sought to question him.

He said at the energy conference that the Obama administration had gone too far with some environmental rules and that he intended to work more closely with industry and individual states to address pollution issues.

"The future ain't what it used to be at the E.P.A.," he said.

Mr. Pruitt's remarks come as the Trump administration prepares to roll back Mr. Obama's two signature policies to address global warming: a pair of sweeping regulations intended to curb carbon dioxide emissions from vehicles and power plant smokestacks.

At the same time, the White House is considering a 17 percent cut to the budget of NOAA, one of the nation's premier agencies of climate science research, according to a memo obtained by The Washington Post.

Mr. Pruitt's remarks on Thursday were consistent with his past public statements questioning the established science of human-caused climate change, but in denying the role played by carbon dioxide, they go a step further.

In addition to putting him at odds with the consensus of climate scientists, Mr. Pruitt's remarks also raise the possibility that, as the Trump administration moves forward with unwinding Mr. Obama's climate change regulations, it could put the administration in violation of federal law.

In 2009, the E.P.A. released a legal opinion known as an endangerment finding concluding that, because of its contribution to global warming, carbon dioxide in large amounts met the Clean Air Act's definition of a pollutant that harms human health. Under the terms of the Clean Air Act, one of the nation's most powerful environmental laws, all such pollutants must be regulated by the E.P.A. A federal court upheld the finding, and the Supreme Court declined to hear a challenge to it.

Thus the E.P.A. remains obligated to regulate carbon dioxide.

In his Senate hearing, Mr. Pruitt said that as administrator of the E.P.A. he would not revisit that 2009 legal finding. "It is there, and it needs to be enforced and respected," Mr. Pruitt said. But energy lobbyists close to the Trump administration have since urged the new administration to consider building a legal case against the endangerment finding.

Advisers to Mr. Trump's transition team said they read Mr. Pruitt's remarks as a signal that he intends to do just that.

"President Trump's campaign commitment was to undo President Obama's entire climate edifice," said Myron Ebell, who worked on Mr. Trump's E.P.A. transition team but has no role in policy making. "They're thinking through the whole thing, and based on what Scott Pruitt said this morning, I do think they are looking at reopening the endangerment finding."

Mr. Trump is expected to announce an executive order next week directing Mr. Pruitt to begin the legal process of unwinding the climate change regulations on emissions from power plants.

JUSTIN GILLIS contributed reporting from New York, and **CLIFFORD KRAUSS** from Houston.

Knowledge, Ignorance and Climate Change

OPINION | BY N. ÁNGEL PINILLOS | NOV. 26, 2018

Philosophers have been talking about skepticism for a long time. Some of those insights can shed light on our public discourse regarding climate change.

NO MATTER HOW SMART or educated you are, what you don't know far surpasses anything you may know. Socrates taught us the virtue of recognizing our limitations. Wisdom, he said, requires possessing a type of humility manifested in an awareness of one's own ignorance. Since then, the value of being aware of our ignorance has been a recurring theme in Western thought: René Descartes said it's necessary to doubt all things to build a solid foundation for science; and Ludwig Wittgenstein, reflecting on the limits of language, said that "the difficulty in philosophy is to say no more than we know."

Awareness of ignorance appears to be common in politics as well. In a recent "60 Minutes" interview, President Trump said of global warming, "I don't know that it's man-made." The same sentiment was echoed by Larry Kudlow, the director of the National Economic Council. Perhaps Trump and Kudlow, confident in their ignorance on these important issues, are simply expressing philosophical humility and wisdom. Or perhaps not.

Sometimes, when it appears that someone is expressing doubt, what he is really doing is recommending a course of action. For example, if I tell you that I don't know whether there is milk in the fridge, I'm not exhibiting philosophical wisdom — I'm simply recommending that you check the fridge before you go shopping. From this perspective, what Trump is doing is telling us that governmental decisions should not assume that global warming is caused by humans.

According to NASA, at least 97 percent of actively publishing climate scientists think that "climate-warming trends over the past century are extremely likely caused by human activities." Americans overwhelmingly agree that the federal government needs to take significant action. In a recent poll conducted by Stanford University, ABC News and Resources for the Future, 61 percent of those surveyed said that the federal government should take a great deal or a lot of action to curb global warming. And an additional 19 percent believe that the government should take moderate action.

As a philosopher, I have nothing to add to the scientific evidence of global warming, but I can tell you how it's possible to get ourselves to sincerely doubt things, despite abundant evidence to the contrary. I also have suggestions about how to fix this.

To understand how it's possible to doubt something despite evidence to the contrary, try some thought experiments. Suppose you observe a shopper at the convenience store buying a lottery ticket. You are aware that the probability that he will lose the lottery is astronomically high, typically above 99.99 percent, but it's hard to get yourself to sincerely say you know this person will lose the lottery. Now imagine your doctor screens you for a disease, and the test comes out negative. But consider the possibility that this result is one of those rare "false negative" cases. Do you really know the result of this particular test is not a false negative?

These scenarios suggest that it's possible to feel as though you don't know something even when possessing enormous evidence in its favor. Philosophers call scenarios like these "skeptical pressure" cases, and they arise in mundane, boring cases that have nothing to do with politics or what one wants to be true. In general, a skeptical pressure case is a thought experiment in which the protagonist has good evidence for something that he or she believes, but the reader is reminded that the protagonist could have made a mistake. If the story is set up in the right way, the reader will be tempted to think that the protagonist's belief isn't genuine knowledge.

When presented with these thought experiments, some philosophy students conclude that what these examples show is that knowledge

DAMON WINTER/THE NEW YORK TIMES

Pump jacks at work bringing up oil in Campbell County, Wy. Climate change skeptics focus on uncertainty even in the face of abundant evidence, N. Ángel Pinillos writes.

requires full-blown certainty. In these skeptical pressure cases, the evidence is overwhelming, but not 100 percent. It's an attractive idea, but it doesn't sit well with the fact that we ordinarily say we know lots of things with much lower probability. For example, I know I will be grading student papers this weekend. Although the chance of this happening is high, it is not anything close to 100 percent, since there is always the chance I'll get sick, or that something more important will come up. In fact, the chance of getting sick and not grading is much higher than the chance of winning the lottery. So how could it be that I know I will be grading and not know that the shopper at the convenience store will lose the lottery?

Philosophers have been studying skeptical pressure intensely for the past 50 years. Although there is no consensus about how it arises, a promising idea defended by the philosopher David Lewis is that skeptical pressure cases often involve focusing on the possibility of error. Once we start worrying and ruminating about this possibility, no matter how far-fetched, something in our brains causes us to doubt. The philosopher Jennifer Nagel aptly calls this type of effect “epistemic anxiety.”

In my own work, I have speculated that an extreme version of this phenomenon is operative in obsessive compulsive disorder, a condition that affects millions of Americans. In many cases of O.C.D., patients are paralyzed with doubt about some fact — against all evidence. For example, a patient might doubt whether she turned off her stove despite having just checked multiple times. As with skeptical pressure cases, the focus on the possibility that one might be wrong plays a central role in the phenomenon.

Let’s return to climate change skepticism. According to social psychology, climate change deniers tend to espouse conservative views, which suggests that party ideology is partly responsible for these attitudes. I think that we should also think about the philosophical nature of skeptical reactions, an apolitical phenomenon.

The standard response by climate skeptics is a lot like our reaction to skeptical pressure cases. Climate skeptics understand that 97 percent of scientists disagree with them, but they focus on the very tiny fraction of holdouts. As in the lottery case, this focus might be enough to sustain their skepticism. We have seen this pattern before. Anti-vaccine proponents, for example, aware that medical professionals disagree with their position, focus on any bit of fringe research that might say otherwise.

Skeptical allure can be gripping. Piling on more evidence does not typically shake you out of it, just as making it even more probable that you will lose the lottery does not all of a sudden make you feel like you know your ticket is a loser.

One way to counter the effects of skepticism is to stop talking about “knowledge” and switch to talking about probabilities. Instead of saying that you don’t know some claim, try to estimate the probability that it is true. As hedge fund managers, economists, policy researchers, doctors and bookmakers have long been aware, the way to make decisions while managing risk is through probabilities. Once we switch to this perspective, claims to "not know," like those made by Trump, lose their force and we are pushed to think more carefully about the existing data and engage in cost-benefit analyses.

Interestingly, people in the grips of skepticism are often still willing to accept the objective probabilities. Think about the lottery case again. Although you find it hard to say you know the shopper will lose the lottery, you readily agree that it is still very probable that he will lose. What this suggests is that even climate skeptics could budge on their estimated likelihood of climate change without renouncing their initial skepticism. It's easy to say you don't know, but it's harder to commit to an actual low probability estimate in the face of overwhelming contrary evidence.

Socrates was correct that awareness of one's ignorance is virtuous, but philosophers have subsequently uncovered many pitfalls associated with claims of ignorance. An appreciation of these issues can help elevate public discourse on important topics, including the future of our planet.

N. ÁNGEL PINILLOS is a professor of philosophy in the School of Historical, Philosophical and Religious Studies at Arizona State University.

CHAPTER 5

Technical Solutions to Climate Crisis

Slashing emissions is an enormous technical challenge, but we're getting closer. Renewable wind and solar power have advanced. Likewise, battery improvements in electric vehicles are changing the car market, and carbon capture technology is being tested on an industrial scale. Beyond high tech, smaller mitigation strategies are emerging in rich and poor nations alike. But such responses will not do much on their own; many argue that only substantial public investment can effectively mobilize these tools and reduce emission levels.

Japan's Growth in Solar Power Falters as Utilities Balk

BY JONATHAN SOBLE | MARCH 3, 2015

MAKURAZAKI, JAPAN — Rice fields, golf courses and even a disused airport runway. All over the southern Japanese region of Kyushu, unexpected places gleam with electricity-producing solar panels.

Solar use in Japan has exploded over the last two years as part of an ambitious national effort to promote renewable energy. But the technology's future role is now in doubt.

Utilities say their infrastructure cannot handle the swelling army of solar entrepreneurs intent on selling their power. And their willingness to invest more money depends heavily on whether the government remains committed to clean energy.

"It's upsetting," said Junji Akagi, a real estate developer on Ukujima, a tiny island near Nagasaki. Mr. Akagi said he hoped to turn a quarter of the island's 10-square-mile area into a "mega-solar" generating station, and has already lined up investors and secured the necessary land.

Then last September, Kyushu Electric Power Company, the region's dominant utility, abruptly announced that it would stop contracting to buy electricity from new solar installations. Other power companies elsewhere in Japan soon followed suit.

"It was a shock," Mr. Akagi said. "Now we don't know if Kyushu Electric will buy our power."

The faltering solar boom is threatening an important goal for Japan as a whole: finding clean sources of power to replace the nuclear output lost after the Fukushima disaster four years ago. So far, the country has been relying mostly on fossil fuels like coal and natural gas to fill the gap, leading to sharply higher emissions of greenhouse gases.

Some solar advocates fear the government is retreating from its clean-energy commitments. Prime Minister Shinzo Abe is pushing to bring back into service some of Japan's 50 nuclear reactors, all of which are now closed as public concern lingers over their safety. If they reopen, it could reduce the need for alternatives like solar power, which many in Mr. Abe's circle, including the powerful industry ministry, see as expensive and unreliable.

Mr. Abe has initiated a review of the renewable-energy policies introduced after Fukushima by a previous, more left-leaning administration. Environmentalists worry he will gut them. The government recently reduced the amount of clean power that utilities are required to buy from outside producers, and additional measures to curb supply are expected this spring, including cuts to price subsidies.

"It would put a brake on the spread of solar power," Yuji Kuroiwa, the ecologically minded governor of Kanagawa Prefecture, next to Tokyo, said at a news conference in December, referring to the new restrictions.

Like other countries that have promoted the technology with generous state support, Japan is also struggling with the financial and technical consequences of its rapid solar growth. Solar power here is costly for consumers because of high state-mandated prices, and handling the fluctuating output of thousands of mostly small solar producers is tricky for utilities. Necessary improvements in the infrastructure have not kept pace, experts say.

"The homework wasn't done," said Nobuo Tanaka, former executive director of the International Energy Agency. Utilities, he said, need to install more hardware — transmission cables, substations and the like — and develop new kinds of expertise to avoid disruptions. "To make renewables work in reality, they have to be properly connected to the power system."

The problem is especially acute in Kyushu, where relatively plentiful sunshine and low land prices have attracted a disproportionate share of solar development.

Installed solar capacity roughly doubled in the two years from mid-2012, when a law took effect requiring utilities to buy renewable energy from outside producers at rates far above market prices. By last summer it stood at 3.4 gigawatts, about equal to the output of three modern nuclear reactors — at least during those hours when the sun was shining at full strength.

More challenging for electric-company planners is what is in the pipeline. An additional 8.4 gigawatts' worth of projects, including Mr. Akagi's on Ukujima, have received government approval but are in limbo after Kyushu Electric's edict. That is more power than the region consumes on some low-demand days — and far too much for Kyushu Electric's grid to handle without the risk of failures, the utility argues.

"If we accept everybody's electricity, the system will become unmanageable," said Shinichi Futami, an official at the utility. It is laying new transmission cables as fast as it can, he added, but has been stymied by the slow, expensive task of securing land rights.

Solar projects have already changed the landscape and economy in Kyushu. They have taken over reservoirs, bankrupt golf courses and idle industrial parks, as well as the more familiar locations of residential rooftops.

The largest ones, like the Nanatsushima Mega-Solar Power Plant in Kagoshima, which opened in 2013, cover areas bigger than 100 football fields. Its vast lot was set aside for a shipyard more than 30 years ago, but sat empty until the recent solar boom.

In Makurazaki, a remote city in Kagoshima, the local airport went unused for a decade, a victim of economic and population decline. Now its 4,200-foot runway is covered end to end with solar panels, a project under the co-ownership of a leasing company and a subsidiary of Kyushu Electric.

The facility is expected to power 2,500 homes this year, its first full year of operation. And instead of paying to maintain an empty airport, Makurazaki will receive about 85 million yen, or a little over $700,000, in annual revenues, mainly from leasing the land.

"The airport was a burden, but now we're getting something for it," said Tadashi Kamizono, the city's mayor.

For all the frantic building, however, Japan still produces less solar power than many other countries. Nationwide, just 2.2 percent of its electricity came from any renewable source in the last fiscal year, excluding hydropower from dams. The small percentage is the legacy of a narrow focus on nuclear power before Fukushima. The figure was less than half the level of the United States or France, and a fraction of the roughly 20 percent achieved by Germany and Spain.

Catching up would be expensive, even if all the necessary infrastructure existed. Japan's financial incentives for solar power and other renewables are the highest in the world — about twice the level of Germany, depending on the type of installation.

According to the government, if every solar plant now on the drawing board were actually to be built, it would cost users ¥2.7

trillion a year in special charges, or about $23 billion, four times the premium they're paying now.

Solar supporters note that the money would at least remain in the Japanese economy, instead of disappearing into the pockets of foreign oil and gas producers. But cost concerns remain.

Higher energy bills related to the nuclear shutdown are already being blamed for squeezing household budgets. That is hurting consumer spending and undercutting Mr. Abe's efforts to jolt the economy to life, through his stimulus program known as Abenomics.

Rather than curtail the expansion of solar power, advocates for the technology say a broader shake-up of Japan's electricity market is needed. Utilities like Kyushu Electric, they argue, have little incentive to accommodate outside competitors to their own coal, gas and nuclear plants. Instead of seeking innovative solutions to the oversupply problem, utilities are using it as an excuse for inaction, said Tomas Kaberger, a Swedish energy expert who heads the Japan Renewable Energy Foundation.

"I presume that it's to protect their economic interests," he said, "not their technological interests or the interests of their customers."

A bill now in Parliament is intended to promote competition. It would force Japan's 10 regional utilities to split their generation and transmission operations into legally separate businesses. The two sides would remain closely connected, however, and some say the plan does not go far enough to even the playing field for new entrants, including those in green energy.

"These 10 monopolies will still own the grid," said Tom O'Sullivan, a Tokyo-based energy consultant. "It will still be very difficult for independent power companies to get their electricity into the grid."

HISAKO UENO contributed reporting.

When Will Electric Cars Go Mainstream? It May Be Sooner Than You Think

BY BRAD PLUMER | JULY 8, 2017

AS THE WORLD'S AUTOMAKERS place larger bets on electric vehicle technology, many industry analysts are debating a key question: How quickly can plug-in cars become mainstream?

The conventional view holds that electric cars will remain a niche product for many years, plagued by high sticker prices and heavily dependent on government subsidies.

But a growing number of analysts now argue that this pessimism is becoming outdated. A new report from Bloomberg New Energy Finance, a research group, suggests that the price of plug-in cars is falling much faster than expected, spurred by cheaper batteries and aggressive policies promoting zero-emission vehicles in China and Europe.

Between 2025 and 2030, the group predicts, plug-in vehicles will become cost competitive with traditional petroleum-powered cars, even without subsidies and even before taking fuel savings into account. Once that happens, mass adoption should quickly follow.

"Our forecast doesn't hinge on countries adopting stringent new fuel standards or climate policies," said Colin McKerracher, the head of advanced transport analysis at Bloomberg New Energy Finance. "It's an economic analysis, looking at what happens when the upfront cost of electric vehicles reaches parity. That's when the real shift occurs."

If that prediction pans out, it will have enormous consequences for the auto industry, oil markets and the world's efforts to slow global warming.

A BOOST FROM BATTERIES

Last year, plug-in vehicles made up less than 1 percent of new passenger vehicle sales worldwide, held back by high upfront costs. The Chevrolet Bolt, produced by General Motors, sells for about $37,500 before federal tax breaks. With gasoline prices hovering around $2 per gallon, relatively few consumers seem interested.

But there are signs of a shift. Tesla and Volkswagen each have plans to produce more than a million electric vehicles per year by 2025. On Wednesday, Volvo announced that it would phase out the traditional combustion engine and that all of its new models starting in 2019 would be either hybrids or entirely battery-powered.

Skeptics argue that these moves are mostly marginal. Exxon Mobil, which is studying the threat that electric cars could pose to its business model, still expects that plug-in vehicle sales will grow slowly, to just 10 percent of new sales in the United States by 2040, with little impact on global oil use. The federal Energy Information Administration projects a similarly sluggish uptick.

The Bloomberg forecast is far more aggressive, projecting that plug-in hybrids and all-electric vehicles will make up 54 percent of new light-duty sales globally by 2040, outselling their combustion engine counterparts.

The reason? Batteries. Since 2010, the average cost of lithium-ion battery packs has plunged by two-thirds, to around $300 per kilowatt-hour. The Bloomberg report sees that falling to $73 by 2030, without any significant technological breakthroughs, as companies like Tesla increase battery production in massive factories, optimize the design of battery packs and improve chemistries.

For the next decade, the report notes, electric cars will remain reliant on government incentives and sales mandates in places like Europe, China and California. But as automakers introduce a greater variety of models and lower costs, electric cars will reach a point where they can stand on their own.

Still, this outcome is hardly guaranteed. Governments could scale back their incentives before plug-in vehicles become fully competitive — many states are already beginning to tax electric cars. Battery manufacturers could face material shortages or production problems that hinder their ability to slash costs. And an unforeseen technology failure, such as widespread battery fires, could halt progress.

"But we tried to be fairly conservative in our estimate of where battery prices are going," Mr. McKerracher said, "and we don't see barriers to electric vehicles' becoming cost competitive very soon."

POTENTIAL SETBACKS

Other experts caution that falling battery costs are not the only factor in determining whether electric cars become widespread. Sam Ori, the executive director of the Energy Policy Institute at the University of Chicago, noted, "People don't buy cars based solely on the price tag."

Consumers may remain wary of vehicles with limited range that can take hours to charge. Even though researchers have shown that battery-electric vehicles have sufficient range for many people's daily commuting habits, consumer psychology is still difficult to predict. The report does not, for instance, expect electric vehicles to catch on widely in the pickup-truck market.

Charging infrastructure is another potential barrier. Although cities are starting to build thousands of public charging stations — and Tesla is working on reducing the time it takes to power a depleted battery — it still takes longer to charge an electric vehicle than it does to refuel a conventional car at the pump.

Many owners charge their cars overnight in their garages, but that is much harder for people living in cities who park their cars on the street.

As a result, the Bloomberg report warns that plug-in vehicles may have a difficult time making inroads in dense urban areas and that infrastructure bottlenecks may slow the growth of electric vehicles after 2040.

Another potential hurdle may be the automakers themselves. While most manufacturers are introducing plug-in models in the United States to comply with stricter fuel-economy standards, they do not always market them aggressively, said Chelsea Sexton, an auto industry consultant who worked on General Motors' electric vehicle program in the 1990s.

Car dealerships also remain reluctant to display and sell electric models, which often require less maintenance and are less profitable for their service departments. Surveys have found that salespeople are often unprepared to pitch the cars.

"We've seen a lot of announcements about electric vehicles, but that doesn't matter much if automakers are just building these cars for compliance and are unenthusiastic about actually marketing them," Ms. Sexton said.

Raw economics may help overcome such barriers, Mr. McKerracher said. He pointed to Norway, where heavy taxes on petroleum-powered vehicles and generous subsidies for electric vehicles have created price parity between the two. As a result, plug-in hybrids and fully electric cars in Norway now make up 37 percent of all new sales, up from 6 percent in 2013.

FIGHTING CLIMATE CHANGE

If Bloomberg's forecast proves correct, it could have sweeping implications for oil markets. The report projects that a sharp rise in electric vehicles would displace eight million barrels of transportation fuel each day. (The world currently consumes around 98 million barrels per day.)

A number of oil companies are now grappling with the prospect of an eventual peak in global demand, with billions of dollars in investments at stake in getting the timing right.

Mass adoption of electric cars could also prove a key strategy in fighting climate change — provided the vehicles are increasingly powered by low-carbon electricity rather than coal. The International Energy Agency has estimated that electric vehicles would have to account for at least 40 percent of passenger vehicle sales by 2040 for the world to have a chance of meeting the climate goals outlined in the Paris agreement, keeping total global warming below 2 degrees Celsius.

Yet the Bloomberg report also shows how much further countries would need to go to cut transportation emissions.

Even with a sharp rise in electric vehicles, the world would still have more traditional petroleum-powered passenger vehicles on the road in 2040 than it does today, and it will take many years to retire existing fleets. And other modes of transportation, like heavy-duty trucking and aviation, will remain stubbornly difficult to electrify without drastic advances in battery technology.

Which means it is still too soon to write an obituary for the internal combustion engine.

On the Attack Against Climate Change

BY ALINA TUGEND | SEPT. 21, 2018

THOUSANDS OF ORGANIZATIONS around the world are trying in big ways and small to confront the challenges of climate change. Here are 10 examples.

UNDER THE SEA

Coral reefs are beautiful to look at, but they also play a crucial role as coastal barriers when storms or flooding hits, absorbing about 97 percent of wave energy.

But because of rising temperatures, coral cover in the Caribbean is estimated to have decreased by about 80 percent in the last few decades, said Joseph Pollock, Caribbean coral strategy director at the Nature Conservancy. He added that in 2016, a marine heat wave was estimated to have killed about a third of the shallow corals on Australia's Great Barrier Reef.

The Nature Conservancy in partnership with Secore International, a conservation organization and a leader in coral restoration, are using an innovative approach to address the problem: helping coral reproduction.

Coral mating works this way, Dr. Pollock said: Many coral species spawn by putting out bundles of eggs and sperm one night a year.

"It's like the craziest singles bar ever," he said.

Researchers know when those nights are, so they go out, collect the eggs and sperm and then mix them together to cross-fertilize, grow them for a few days or weeks until they become coral juveniles, then place them back in the sea.

The survival rate is about 10 percent, Dr. Pollock said, but that's much better than the survival rate without the help of the scientists. And compared with other restoration techniques, the cross-fertilization creates greater genetic diversity, and that creates more resilience.

The work is focused on the Caribbean now, but Dr. Pollock said the hope is that it can be used throughout the world.

"The aim of the work is to develop tools and techniques that are low cost and don't require a huge amount of super specialized personnel and infrastructure," he said.

A NEW KIND OF POWER

After Hurricane Maria swept through Puerto Rico last year and knocked out almost all of its power grid, most residents were left without electricity for months. Jonathan Marvel, one of the founders of Marvel Architects, was born and raised in Puerto Rico. He wanted not only to help bring back power but also do it in a way that would be more environmentally sustainable than it was before the storm.

So, with colleagues and friends, he created Resilient Power Puerto Rico, to develop and install solar microgrids for as many people as he could in the most efficient way possible. The organization, which received an early donation of batteries from Tesla, focused its efforts on areas with high-density, low-rise housing and installed the grids on rooftops of community centers that typically serve 3,000 to 4,000 people.

Their work complements other efforts not only to rebuild the island but also to make its infrastructure more resilient and environmentally green.

One benefit of the solar microgrids is that they can store solar power — allowing them to operate if the main power source is disrupted — which solar panels alone can't do.

So far, 28 microgrids have been installed, serving close to 100,000 people, Mr. Marvel said, and 30 more are almost finished. The cost, covered by donations from companies and individuals, is about $25,000 to $30,000 per solar hub.

Almost all of the power on the island is supplied by fossil fuels, but Puerto Rico is "an ideal locale to use solar power and renewable energy because it has so many more solar days than in many parts of the world," said Mr. Marvel, whose offices are based in Manhattan and

San Juan. "We want to keep the candle burning with solar energy, not fossil fuel."

SAVING THE SOIL

The role that soil plays in climate change is often ignored, but changing the way it is managed could have a big impact on global warming.

Unfortunately, most soil has become less productive, with environmental consequences, said Michael Doane, managing director for agriculture and food systems at the Nature Conservancy. That's because it has been eroded through too much tilling, lack of adequate ground cover and a failure to diversify crops.

"This living ecosystem has become dead and we're trying to bring it back to life," Mr. Doane said.

One pilot program, now taking place on more than 100 American farms in about six states, is focused on reducing or eliminating the amount of tillage done on farms. It is done under the auspices of the Soil Health Partnership, a collaboration of environmental groups, farmers, academics and industry working to alter soil health practices.

"Tillage is actually detrimental to soil," he said. One of the main problems is that tilling releases carbon stored in the soil, which becomes carbon dioxide when exposed to air and contributes to global warming. Tilling also makes the earth more susceptible to erosion and less able to absorb heavy rainfalls.

One solution is using plants — either rotating crops or using ground cover such as grass, depending on what's needed to repair the soil — to cover the soil before and after the main cash crop is planted. Diverse plant cover has been found to make the soil healthier and helps control weeds, Mr. Doane said.

"We want to try to avoid soil bare of plant cover," he added. "Instead, our vision is a continuous living cover." Calling it "nature's solution to climate change," he said the process of photosynthesis — where plants store the carbon in the soil and release oxygen — could be "a very cost-effective way to mitigate climate change."

This won't work on every farm, because each is different, but "we know this works for many farmers in many situations — we have good data on that," Mr. Doane said.

And the process can make farms more productive by creating soil that can better hold water and recycle nutrients, meaning farmers can spend less money on fertilizer.

"If we're going to solve climate change," he said, "We have to find economic solutions for people who don't know they're solving climate change."

UP ON THE ROOF

Keeping cool is becoming more and more difficult as temperatures across the world spike. In addition, air-conditioning uses hydrofluorocarbons, which contribute substantially to global warming.

One solution, which numerous cities around the world have embraced, is called cool roofs, which is simply painting dark rooftops with a reflective white paint or wrapping them with a light membrane that reduces the absorption of heat. This not only addresses the "urban heat island" effect — urban areas tend to be significantly warmer than surrounding rural areas due to human activities — but also helps decrease strain on electric grids and alleviate air pollution.

A Yale University study cites a finding that if every roof in the United States were painted white, the urban heat island effect would be decreased by one-third.

New York City, for example, has a CoolRoofs Initiative. Since 2009, 5,000 volunteers have painted more than five million square feet of rooftops in the city, according to the Mayor's Office of Sustainability.

In India, where only 10 percent of the households have air-conditioning units, two cities — Hyderabad and Ahmedabad — ran cool-roof demonstration projects. In Ahmedabad, volunteers and others painted 3,000 roofs in slum areas with white lime paint, said Anjali Jaiswal, director of the Natural Resources Defense Council's

India program. The environmental organization worked on the projects with local partners.

DuPont, which has a research center in Hyderabad, owns Tyvek, a synthetic material that is often used in construction and can cover dark roofs. The company donated the material to cover 25 roofs in the city. Both paint and the coverings are considered equally effective, Ms. Jaiswal said.

Cool roofs can reduce indoor temperature by three to nine degrees Fahrenheit, she added, for as little as seven cents a square foot or $4 a home.

As incomes and temperatures rise, so is demand for air-conditioning, she said. An important aspect of addressing climate change will be both developing more environmentally friendly units and reducing the demand for them.

Both Indian cities are now developing a citywide cool-roofs policy mandating them for all municipal buildings and working with business leaders' corporate responsibility programs to expand them throughout the cities.

For relatively little cost, Ms. Jaiswal said, cool roofs are "saving lives, reducing temperatures and responding to climate change."

TURNING PLASTIC INTO MONEY

Collecting plastic to recycle as a way to earn money is nothing new. But David Katz, founder and chief executive of the Plastic Bank, has created a virtuous cycle of buying and reselling the plastic.

The company's idea, which last year received one of the United Nations' "Momentum for Change" climate solutions awards, aims to stop plastic before it even gets to the ocean by having collectors pick it up around canals, waterways and other areas that lead into the ocean. Through partnerships between the Plastic Bank and major corporations such as the German-based Henkel, the plastic is then reused. That cuts down on the emissions that cause greenhouse gasses used to make new plastic.

According to research from the nonprofit Earth Day Network, about eight million metric tons of plastic pollution are discarded into the ocean every year, equivalent to one garbage truck full of plastics being dumped every minute.

Plastic Bank, based in Vancouver, British Columbia, started its work in Haiti three years ago; now about 2,000 collectors there can either receive cash, buy goods or services — such as cooking oil, LED lights or topping off pay-as-you-go cellphones — at one of the 40 recycling outlets around the country. They also have the ability, through the Plastic Bank's app developed in partnership with IBM, to transfer the money into an online savings account.

In Haiti, where more than half the people live on less than $2 a day, a full-time collector can receive several dollars a day, Mr. Katz said.

The company also trains and supports local people who run the recycling outlets.

The Plastic Bank has expanded to Brazil, and Indonesia; this month, it opened its first site in the Philippines and in the first week, Mr. Katz said, collected around 120,000 bottles.

NEIGHBORS HELPING NEIGHBORS

Constance Okollet had never heard of climate change, but she knew that her village in Uganda had been devastated by a 2007 flood that affected most of the country. She knew the weather was growing increasingly unpredictable, making the farming of the typical crops such as maize, sorghum and millet ever more difficult and sending a population that had been poor but self-sufficient spiraling into destitution.

"We thought God was punishing us," Ms. Okollet said. She suggested to her neighbors that they form a group to help one another and was elected to lead what was soon called the Osukuru United Women Network.

At first, they helped each other in small ways, such as pooling their savings. Then in 2009, Ms. Okollet heard on the radio that Oxfam, the global relief organization, was holding a meeting focused on food insecurity in the area, and she decided to go.

Once at the meeting, she said: "They kept talking about climate change and I asked, 'What are you people talking about? What do you mean by that?' "

She learned. And she and other members of the network (which now includes some men) have since begun awareness education about climate change — its impact and how to adapt — through workshops in churches and wherever people gather.

They have undertaken numerous larger projects as well. The network received a $5,000 grant from the Global Greengrants Fund, a nonprofit that provides small grants to local groups working on environmental issues. The money went to buy six teams of oxen, which are much faster than the traditional hand tilling. An acre can be tilled in two days, compared with a hoe, which can take four weeks. Ms. Okollet said. This makes it easier to time the planting to good weather.

Two years ago, 60 members of the network were also flown to Nairobi, Kenya, to learn how to make and sell charcoal briquettes; deforestation means firewood is scarce and the briquettes in any case are greener. They mix ash, dry leaves and water, which when dried, actually even cook better than wood, she said.

"We also sell the briquettes to make money — even $1 can help," she said. "You can pay your school fees or start a small business, and you don't have to take a loan from the bank."

RESTORING PEATLANDS

Peatlands may not be the first thing people think about when focusing on climate change, but the abandonment of drained peatlands in parts of Russia has created not only widespread land degradation, but also huge quantities of carbon dioxide, through peat oxidation. And carbon dioxide contributes to global warming.

Over the decades, millions of acres have been drained and used for agriculture, forestry and the extraction of peat, a fuel used for heating and electrical energy. But when it was no longer profitable to dig out the peat, many of the areas were deserted, said Jozef Bednar, project manager for Wetlands International.

"Peatland ecosystems play a crucial role in global climate," said Dr. Bednar, noting that they store several times more carbon dioxide, the leading greenhouse gas than any other ecosystem. As such, he added, "the world's peat bogs represent an important 'carbon sink' — a place where carbon dioxide is stored below ground and can't escape into the atmosphere and exacerbate global warming."

Dr. Bednar offered one staggering number: Peatlands cover only 3 percent of the global total land area, but emit twice as much carbon dioxide as the world's forests, which cover more than 30 percent. The peatlands drained by people are prone to fires and the accompanying smoke spreads long distances, creating serious health problems.

Wetlands International, along with its partners under the International Climate Initiative of the German government, began a major restoration of the peatlands after the extensive peat fires in the Moscow region in 2010. The goal is to return the peatlands to their original waterlogged state. With the help of experts, this is done by correctly blocking drainage ditches and channels so the peatlands' water-storage capacity is re-established, Dr. Bednar said.

The project was awarded a United Nation's "Momentum for Change" climate solutions award last year and, to date, about 100,000 acres of drained peatlands have been restored in Russia and the process can be replicated in other countries facing the same problem, he added.

CLIMATE LITERACY

Climate and climate change are complicated, and while schools are a good place to learn about it, not all teachers have the knowledge and resources to teach the topic. That's why the United States National Oceanic and Atmospheric Administration and a partnership of federal agencies, education-focused nongovernmental organizations, teachers and scientists wrote "The Essential Principles of Climate Literacy," a curriculum guide for teachers.

Available since 2009, but in the process of being updated for release at the end of the year, it is for at all ages and all forms of education, said Frank Niepold, a senior climate education program manager for NOAA and lead author of the guide.

"In the 1990s, less than 1 percent of the national standards for science education was related to climate change. Now about 30 percent is," he added.

At the same time, the partnership established a website, Cleanet.org, that offers climate and energy educational resources — including quizzes — and guidance for teachers.

Mr. Niepold estimated that over 50 percent of children in kindergarten through 12th grade nationwide are learning from all or some of the climate literacy framework, and "we're on our way to 75 percent," he said.

Other countries are also using the guide in creating their own curriculums and standards, he added, and this month the National Science Teachers Association released its position paper on teaching climate science, referencing the Essentials of Climate Literacy as one of its sources.

"Students are aware of climate change and want to know more and want to be part of solving it," Mr. Niepold said. "And they know that requires an understanding of the fundamentals."

FIGHTING ENERGY POVERTY

When Laura Stachel, an obstetrician, took a two-week trip to a remote hospital in Nigeria 10 years ago, she was interested in maternal health, not solar energy. But what she saw there changed her mind and her life.

She knew maternal mortality was high: Worldwide, Dr. Stachel said, about 300,000 women and two million newborns die every year from pregnancy and childbirth complications. But she did not realize the extent of what has been called energy poverty.

The hospital in northern Nigeria that she visited did not have electricity for 12 hours a day. Daytime cesarean sections were done by ambient light, and once, when it occurred in the middle of the night and the power went out, one was performed by the light of Dr. Stachel's headlamp.

She told the stories to her husband, Hal Aronson, who holds a doctorate in environmental sociology and has focused on solar energy issues for years. He designed and built what is now called a Solar Suitcase: solar equipment that is easy to transport, install and use in areas where power supplies are unreliable.

The kit, the size of a suitcase, comes with everything needed from solar panels to medical lighting to fetal monitors. As news about the Solar Suitcases was spreading, Dr. Stachel and Dr. Aronson also started the nonprofit WeCareSolar, which has received grants from foundations, corporations and individuals. Last year, it received a United Nations "Momentum for Change" climate solutions award.

Working in partnership with nonprofits and United Nations agencies, about 3,500 facilities in 27 developing countries around the world have received the Solar Suitcases. It costs $3,000 to support a clinic with a Solar Suitcase for five years, Dr. Stachel said, including all the equipment, transportation and training.

The organization also works to train local people to install and maintain solar power.

The health impact is clear, but so is the impact on the environment, she said. Diesel fuel generators and kerosene lamps are polluting and generate carbon dioxide. But perhaps even more important, the move toward solar would reduce the reliance on fossil fuel — something that some major American hospitals are now trying to do.

"We could leapfrog right past that and go right to clean, green electricity," she said.

GREENER REFRIGERATION

Supermarkets around the world are major users of hydrofluorocarbon refrigerants, which contribute to ozone depletion and global warming — and in Chile, they are the biggest user. So, it is fitting that a supermarket chain called Jumbo has become the first in the country to adopt new refrigeration technology that is far more climate friendly than traditional methods.

The new refrigeration technology uses transcritical CO2, which is a refrigerant that has a much smaller effect on the ozone layer and global warming. Hydrofluorocarbon refrigerants had replaced ozone-depleting chlorofluorocarbons, but because their effects on global warming are so severe, there has been a worldwide effort to find a replacement. Hydrofluorocarbons have 1,000 times the heat-trapping potency of carbon dioxide.

Under the Montreal Protocol's Kigali Amendment, countries must meet specific targets and timelines to replace hydrofluorocarbon refrigerants with more environmentally friendly alternatives.

So far, Jumbo has installed the systems in three supermarkets in Chile and will convert four more stores in the near future, said Claudia Paratori Cortés, coordinator of the Ozone Unit in in the Office of Climate Change in Chile's Ministry of Environment.

Ms. Cortés said comparisons between two types of refrigeration — transcritical CO2 and one containing hydrofluorocarbons — found that the transcritical CO2 systems were 20 percent to 40 percent more energy efficient, saving around $20,000 annually.

In addition, she said, the residual heat from the transcritical CO2 systems can be used to heat water and therefore save energy.

An Energy Wish List for Congress

OPINION | BY JUSTIN GILLIS AND HAL HARVEY | FEB. 5, 2019

AMID THE DISARRAY in Washington, here is a ray of hope: It seems possible that Congress could pass energy legislation this year with the support of both parties. It would not be the sweeping measure to tackle climate change that is really needed, but there is at least a chance of getting a bill that does more good than harm for the climate and the country.

The point person on this effort will be Lisa Murkowski, the maverick Republican from Alaska who heads the Senate Energy and Natural Resources Committee. She has a history of working with Democrats on energy legislation, and got 85 Senate votes for passage of a major energy bill in 2016, only to see it bog down in negotiations with a House then also led by Republicans.

Newly installed as leader of the Democrats on her committee is Joe Manchin, of West Virginia. Both senators are beholden to fossil-fuel interests, and it is inevitable that any bill they draft will seek to include provisions sought by that powerful lobby. But Ms. Murkowski has seen the effects of climate change firsthand in Alaska, and takes the problem seriously. Mr. Manchin declared the other day, "We want clean water; we want clean air; we want to have an economy that works in balance with the environment, we really do."

In the House, newly empowered Democrats are eager to tackle climate change, but as long as Mitch McConnell, the Senate majority leader, controls the Senate calendar, getting a sweeping climate bill through that body will be impossible. So, even as Democrats work to take control of the Senate in the 2020 election, they ought to engage with Ms. Murkowski to try to get a reasonable measure out of Congress — and to keep it focused on clean energy.

In a promising beginning, the energy committee is scheduled to hold a hearing on Thursday to examine the status and outlook for energy innovation. What, then, ought to be in this year's energy bill? Here are some ideas that should appeal to both parties.

First, Senators Murkowski and Manchin should try to establish a firm consensus in Congress that the nation must, at minimum, double its annual spending on energy research and development. They could establish this intent in the legislation, then push as a committee for the appropriations. Business and scientific leaders, including Bill Gates, have been pleading with Congress to triple the budget for years. The sum being spent for both basic and applied research now, less than $6 billion a year in the Department of Energy budget, is paltry compared to the scope of our energy and climate problems. The United States risks falling behind China and other countries in energy technology.

In theory, Congress agreed to double the nation's overall research budget, including energy research, in a 2007 law, then reiterated that commitment in 2010. But it never actually put up the money. The House can play a major role in ensuring that this broken promise is finally kept. At a bare minimum, it should insist that an extra $1 billion to $2 billion a year be allocated for applied clean energy research.

Second, the federal tax breaks that support installation of renewable energy technologies like wind turbines and solar panels should be extended for five years. Congress should also lift caps on federal tax credits for electric cars. Under current law, these incentives will be phased out in the next few years; they should go eventually, but not until the technologies are more widespread.

Ms. Murkowski and Mr. Manchin should specifically seek a rapid expansion of offshore wind farms. That would require tax breaks to jump-start that industry, as well as a mandate that the Trump administration accelerate offshore leasing. The technology of offshore wind production, developed largely in Europe, has improved significantly in recent years; immense turbines, planted far from shore, can each supply electricity for up to 8,000 homes. States like New York and Massachusetts are already making big plans for offshore wind, but Congress and the president could speed things up. As the market expands, costs will fall and the tax breaks can be phased out.

Third, Congress ought to make it easier to build new high-capacity power lines across the country. This sounds rather technical, we know, but it is vital. As wind turbines and solar farms become a bigger part of the electricity mix, one of the best ways to offset their intermittent nature will be to transmit power over longer distances — if the wind is not blowing in Kansas, it may well be blowing in Oklahoma or North Dakota. But state parochialism is slowing the construction of the needed power lines.

Congress actually tried to solve this problem in 2005, but the language it wrote then was ambiguous. The courts essentially gutted the relevant part of the law and gave the states a veto power over new transmission line projects. Congress needs to grant the main federal power agency, the Federal Energy Regulatory Commission, clear authority to approve a new generation of power corridors. The agency must consult with the states and the public, and respect environmental laws, of course, but if a power corridor is in the broad national interest, it ought to get built.

These new power corridors would be an investment in rural America. Wind turbines can increase farmers' incomes while using only a small portion of their land, and that income won't crash when commodity prices do. Across the windy middle of the country, farmers have come to see the wind as a crop they can harvest all year — and it is less work than growing corn. But a lack of transmission capacity has slowed some projects. Why not get serious about taking this product to market and allow farmers to become the core suppliers of electricity for the next several generations?

Congress should also tackle a related problem. New electronic devices that monitor the condition of long-distance power lines can help to increase their effective capacity at little cost. But utilities get paid for building new equipment, not for using old equipment more efficiently, so they are dragging their feet about installing these upgrades. The federal government ought to mandate it on interstate power lines, and set a deadline for compliance.

CHANG W. LEE/THE NEW YORK TIMES

Funding to encourage the rapid expansion of offshore wind farms would be a promising beginning for climate legislation.

Senator Murkowski's 2016 bill had a host of energy-efficiency programs that would ultimately have saved consumers tens of billions of dollars every year; these measures should be included in the new bill.

Finally, we urge Mr. Manchin and other coal-state senators to recognize that the coal industry has gone into long-term decline; coal is simply losing market share to natural gas and renewable energy. The fair thing to do is to fund a transition package for coal miners and their communities, conceptually similar to what Congress did when it passed a buyout program for tobacco farmers when a federal support policy ended.

These provisions will not fully satisfy the wish lists of either party. But they would speed the development of the nation's clean-energy economy. Wouldn't it be nice, amid the daily sniping in the nation's Capitol, to show the world that America can still solve big problems?

JUSTIN GILLIS, a former environmental reporter for The Times, is a contributing opinion writer. **HAL HARVEY** is the chief executive of the research firm Energy Innovation.

The Tiny Swiss Company That Thinks It Can Help Stop Climate Change

BY JON GERTNER | FEB. 12, 2019

Two European entrepreneurs want to remove carbon from the air at prices cheap enough to matter.

JUST OVER A CENTURY AGO in Ludwigshafen, Germany, a scientist named Carl Bosch assembled a team of engineers to exploit a new technique in chemistry. A year earlier, another German chemist, Fritz Haber, hit upon a process to pull nitrogen (N) from the air and combine it with hydrogen (H) to produce tiny amounts of ammonia (NH3). But Haber's process was delicate, requiring the maintenance of high temperatures and high pressure. Bosch wanted to figure out how to adapt Haber's discovery for commercial purposes — as we would say today, to "scale it up." Anyone looking at the state of manufacturing in Europe around 1910, Bosch observed, could see that the task was daunting: The technology simply didn't exist.

Over the next decade, however, Bosch and his team overcame a multitude of technological and metallurgical challenges. He chronicled them in his 1932 acceptance speech for the Nobel Prize for Chemistry — an honor he won because the Haber-Bosch process, as it came to be known, changed the world. His breakthrough made possible the production of ammonia on an industrial scale, providing the world with cheap and abundant fertilizer. The scientist and historian Vaclav Smil called Haber-Bosch "the most important technical invention of the 20th century." Bosch had effectively removed the historical bounds on crop yields, so much so that he was widely credited with making "bread from air." By some estimates, Bosch's work made possible the lives of more than two billion human beings over the last 100 years.

What the Haber-Bosch method had going for it, from the very start, was a ready market. Fertilizer was already in high demand, but it came primarily from limited natural reserves in far-flung locales — bird droppings scraped from remote islands near Peru, for instance, or mineral stores of nitrogen dug out of the Chilean desert. Because synthetic ammonia competed with existing products, it was able to follow a timeworn pattern of innovation. In much the same way that LEDs have supplanted fluorescent and incandescent bulbs (which in turn had displaced kerosene lamps and wax candles), a novel product or process often replaces something already in demand. If it is better or cheaper — and especially if it is better *and* cheaper — it usually wins in the marketplace. Haber-Bosch did exactly that.

It may now be that another gas — carbon dioxide (CO2) — can be removed from the air for commercial purposes, and that its removal could have a profound effect on the future of humanity. But it's almost certainly too soon to say for sure. One sunny morning last October, several engineers from a Swiss firm called Climeworks ambled onto the roof of a power-generating waste-incineration plant in Hinwil, a village about 30 minutes outside Zurich. The technicians had in front of them 12 large devices, stacked in two rows of six, that resembled oversize front-loading clothes dryers. These were "direct air capture" machines, which soon would begin collecting carbon dioxide from air drawn in through their central ducts. Once trapped, the CO2 would then be siphoned into large tanks and trucked to a local Coca-Cola bottler, where it would become the fizz in a soft drink.

The machines themselves require a significant amount of energy. They depend on electric fans to pull air into the ducts and over a special material, known as a sorbent, laced with granules that chemically bind with CO2; periodic blasts of heat then release the captured gas from the sorbent, with customized software managing the whole catch-and-release cycle. Climeworks had installed the machines on the roof of the power plant to tap into the plant's low-carbon electricity and the heat from its incineration system. A few dozen yards away from the new installation sat an older stack of Climeworks machines, 18 in total, that had been whirring on the same rooftop for more than a year. So far, these machines had captured about 1,000 metric tons (or about 1,100 short tons) of carbon dioxide from the air and fed it, by pipeline, to an enormous greenhouse nearby, where it was plumping up tomatoes, eggplants and mâche. During a tour of the greenhouse, Paul Ruser, the manager, suggested I taste the results. "Here, try one," he said, handing me a crisp, ripe cucumber he plucked from a nearby vine. It was the finest direct-air-capture cucumber I'd ever had.

Climeworks's rooftop plant represents something new in the world: the first direct-air-capture venture in history seeking to sell CO2 by the ton. When the company's founders, Christoph Gebald and Jan Wurzbacher, began openly discussing their plans to build a business several years ago, they faced a deluge of skepticism. "I would say nine out of 10 people reacted critically," Gebald told me. "The first thing they said was: 'This will never work technically.' And finally in 2017 we convinced them it works technically, since we built the big plant in Hinwil. But once we convinced them that it works technically, they would say, 'Well, it will never work economically.' "

For the moment, skeptics of Climeworks's business plan are correct: The company is not turning a profit. To build and install the 18 units at Hinwil, hand-assembled in a second-floor workshop in Zurich, cost between $3 million and $4 million, which is the primary reason it costs the firm between $500 and $600 to remove a metric ton of CO2 from the air. Even as the company has attracted about $50 million in private investments and grants, it faces the same daunting task that confronted Carl Bosch a century ago: How much can it bring costs down? And how fast can it scale up?

Gebald and Wurzbacher believe the way to gain a commercial foothold is to sell their expensive CO2 to agriculture or beverage companies. Not only do these companies require CO2 anyway, some also seem willing to pay a premium for a vital ingredient they can use to help market their products as eco-friendly.

Still, greenhouses and soda bubbles together represent a small global market — perhaps six million metric tons of CO2 annually. And Gebald and Wurzbacher did not get into carbon capture to grow mâche or put bubbles in Fanta. They believe that over the next seven years they can bring expenses down to a level that would enable them to sell CO2 into more lucrative markets. Air-captured CO2 can be combined with hydrogen and then fashioned into any kind of fossil-fuel substitute you want. Instead of making bread from air, you can make fuels from air. Already, Climeworks and another company, Carbon Engineering, which is based in British Columbia, have moved aggressively on this idea; the Canadians have even lined up investors (including Bill Gates) to produce synthetic fuel at large industrial plants from air-captured CO2.

The ultimate goal for air capture, however, isn't to turn it into a product — at least not in the traditional sense. What Gebald and Wurzbacher really want to do is to pull vast amounts of CO2 out of the atmosphere and bury it, forever, deep underground, and sell that service as an offset. Climeworks's captured CO2 has already been injected deep into rock formations beneath Iceland; by the end of the year, the firm intends to deploy 50 units near Reykjavik to expand the operation. But at that point the company will be moving into uncharted economic territory — purveyors of a service that seems desperately needed to help slow climate change but does not, at present, replace anything on the consumer or industrial landscape. To complicate matters, a ton of buried CO2 is not something that human beings or governments have shown much demand for. And so companies like Climeworks face a quandary: How do you sell something that never existed before, something that may never be cheap, into a market that is not yet real?

Even the most enthusiastic believers in direct air capture stop short of describing it as a miracle technology. It's more frequently described as an old idea — "scrubbers" that remove CO2 have been used in submarines since at least the 1950s — that is being radically upgraded for a variety of new applications. It's arguably the case, in fact, that when it comes to reducing our carbon emissions, direct air capture will be seen as an option that's too expensive and too modest in impact. "The only way that direct air capture becomes meaningful is if we do all the other things we need to do promptly," Hal Harvey, a California energy analyst who studies climate-friendly technologies and policies, told me recently. Harvey and others make the case that the biggest, fastest and cheapest gains in addressing atmospheric carbon will come from switching our power grid to renewable energy or low-carbon electricity; from transitioning to electric vehicles and imposing stricter mileage regulations on gas-powered cars and trucks; and from requiring more energy-efficient buildings and appliances. In short, the best way to start making progress toward a decarbonized world is not to rev up millions of air capture machines right now. It's to stop putting CO2 in the atmosphere in the first place.

The future of carbon mitigation, however, is on a countdown timer, as atmospheric CO2 concentrations have continued to rise. If the nations of the world were to continue on the current track, it would be impossible to meet the objectives of the 2016 Paris Agreement, which set a goal limiting warming to 2 degrees Celsius or, ideally, 1.5 degrees. And it would usher in a world of misery and economic hardship. Already, temperatures in some regions have climbed more than 1 degree Celsius, as a report by the Intergovernmental Panel on Climate Change noted last October. These temperature increases have led to an increase in droughts, heat waves, floods and biodiversity losses and make the chaos of 2 or 3 degrees' additional warming seem inconceivable. A further problem is that maintaining today's emissions path for too long runs the risk of doing irreparable damage to the earth's ecosystems — causing harm that no amount of technological innovation can make right. "There is no reverse gear for natural systems," Harvey says. "If they go, they go. If we defrost the tundra, it's game over." The same might be said for the Greenland and West Antarctic ice sheets, or our coral reefs. Such resources have an asymmetry in their natural architectures: They can take thousands or millions of years to form, but could reach conditions of catastrophic decline in just a few decades.

At the moment, global CO2 emissions are about 37 billion metric tons per year, and we're on track to raise temperatures by 3 degrees Celsius by 2100. To have a shot at maintaining a climate suitable for humans, the world's nations most likely have to reduce CO2 emissions drastically from the current level — to perhaps 15 billion or 20 billion metric tons per year by 2030; then, through some kind of unprecedented political and industrial effort, we need to bring carbon emissions to zero by around 2050. In this context, Climeworks's effort to collect 1,000 metric tons of CO2 on a rooftop near Zurich might seem like bailing out the ocean one bucket at a time. Conceptually, however, it's important. Last year's I.P.C.C. report noted that it may be impossible to limit warming to 1.5 degrees by 2100 through only a rapid switch to clean energy, electric cars and the like. To preserve a livable environment we may also need to extract CO2 from the atmosphere. As Wurzbacher put it, "if you take all these numbers from the I.P.C.C., you end up with something like eight to 10 billion tons — gigatons — of CO2 that need to be removed from the air every year, if we are serious about 1.5 or 2 degrees."

There happens to be a name for things that can do this kind of extraction work: negative-emissions technologies, or NETs. Some NETs, like trees and plants, predate us and probably don't deserve the label. Through photosynthesis, our forests take extraordinary amounts of carbon dioxide from the atmosphere, and if we were to magnify efforts to reforest clear-cut areas — or plant new groves, a process known as afforestation — we could absorb billions more metric tons of carbon in future years. What's more, we could grow crops specifically to absorb CO2 and then burn them for power generation, with the intention of capturing the power-plant emissions and pumping them underground, a process known as bioenergy with carbon capture and storage, or BECCS. Other negative-emissions technologies include manipulating farmland soil or coastal wetlands so they will trap more atmospheric carbon and grinding up mineral formations so they will absorb CO2 more readily, a process known as "enhanced weathering."

Negative emissions can be thought of as a form of time travel. Ever since the Industrial Revolution, human societies have produced an excess of CO2, by taking carbon stores from deep inside the earth — in the form of coal, oil and gas — and from stores aboveground (mostly wood), then putting it into the atmosphere by burning it. It has become imperative to reverse the process — that is, take CO2 out of the air and either restore it deep inside the earth or contain it within new surface ecosystems. This is certainly easier to prescribe than achieve. "All of negative emission is hard — even afforestation or reforestation," Sally Benson, a professor of energy-resources engineering at Stanford, explains. "It's not about saying, 'I want to plant a tree.' It's about saying, 'We want to plant a billion trees.' " Nevertheless, such practices offer a glimmer of hope for meeting future emissions targets. "We have to come to grips with the fact that we waited too long and that we took some options off the table," Michael Oppenheimer, a Princeton scientist who studies climate and policy, told me. As a result, NETs no longer seem to be just interesting ideas; they look like necessities. And as it happens, the Climeworks machines on the rooftop do the work each year of about 36,000 trees.

Last fall, the National Academies of Sciences, Engineering and Medicine published a lengthy study on carbon removal. Stephen Pacala, a Princeton professor who led the authors, pointed out to me that negative-emissions technologies have various strengths and drawbacks, and that a "portfolio" approach — pursue them all, then see which are the best — may be the shrewdest bet. If costs for direct air capture can be reduced, Pacala says he sees great promise, especially if the machines can offset emissions from economic sectors that for technological reasons will transition to zero carbon much more slowly than others. Commercial aviation, for instance, won't be converted to running on solar power anytime soon. Jennifer Wilcox, a chemical-engineering professor at Worcester Polytechnic Institute, in Massachusetts, told me that air capture could likewise help counter the impact of several vital industries. "There are process emissions that come from producing iron and steel, cement and glass," she says, "and any time you make these materials, there's a chemical reaction that emits CO2." Direct air capture could even lessen the impacts of the Haber-Bosch processes for making fertilizer; by some estimates, that industry now accounts for 3 percent of all CO_2 emissions.

Pacala equates the challenges confronting Climeworks and Carbon Engineering to what the wind- and solar-power industries faced in the 1970s and '80s, when their products were expensive compared with fossil fuels. Those industries couldn't rely on demand from the private sector alone. But some policymakers perceived tremendous environmental and public benefits if they could surmount that hurdle. Government investments in research, along with state and federal tax credits, helped the young industries expand. "Wind and solar are now the cheapest forms of energy in the right locations," Pacala says. "The return on those investments, if you calculated it, would blow the doors off anything in your portfolio. It's like investing in early Apple. So it's a spectacular story of success. And direct air capture is precisely the same kind of problem, in which the only barrier is that it's too costly."

Most of Climeworks's 60 employees work in a big industrial space in downtown Zurich, on two floors of a low-slung building that the company sublets from a German aerospace firm. Manufacturing operations are on the ground floor; the research labs are upstairs, along with a small suite of shared offices, a hallway kitchen and a hangout area. The place has the stark, casual feel of a tech start-up, with one exception: The walls are lined with oversize photos of pivotal moments in Climeworks's young history — its ungainly early prototypes; the opening of the first Hinwil plant that collected CO_2 for the greenhouse.

"It's a little bit by accident that we are based in Switzerland," Wurzbacher told me. He and Gebald both grew up in Germany and met as undergraduates at E.T.H. Zurich, the Swiss Federal Institute of Technology. "We met on Day 1, on the 20th of October of 2003," Gebald recalled. "And on Day 1 we decided that we'd have a company." Their aspiration was to be entrepreneurs, not to start a carbon-capture firm, but both men were drawn to research on renewable energy and reducing emissions. After they completed their master's projects, they decided to create a direct-air-capture prototype and go into business.

Both took the title of company director. Helped by a number of small grants, Climeworks was incorporated in 2009.

The two men were not alone in trying to chip away at decades of carbon emissions. An American start-up, Global Thermostat, now finishing its first commercial plant in Alabama, began working on air-capture machines in 2010. And almost from the start, Gebald and Wurzbacher found themselves in a friendly competition with David Keith, the Harvard engineering professor who had just started Carbon Engineering in British Columbia. Keith's company settled on a different air-capture technology — employing a higher-heat process, and a liquid solution to capture CO2 — to brew synthetic fuels. Climeworks's big advantage is that it can make smaller plants early, Keith told me: "I am crazy jealous. It's because they're using a modular design, and we're not." On the other hand, Keith said he believes his firm is closer to building a big plant that could capture carbon at a more reasonable cost and produce substantial amounts of fuel. "I don't see a path for them to match this." Gebald told me he thinks his and Keith's companies will each succeed with differing approaches. For now, what all the founders have in common is a belief that the cost of capturing a ton of carbon will soon drop sharply.

Their view is not always shared by outside observers. M.I.T.'s Howard Herzog, for instance, an engineer who has spent years looking at the potential for these machines, told me that he thinks the costs will remain between $600 and $1,000 per metric ton. Some of Herzog's reasons for skepticism are highly technical and relate to the physics of separating gases. Some are more easily grasped. He points out that because direct-air-capture machines have to move tremendous amounts of air through a filter or solution to glean a ton of CO2 — the gas, for all its global impact, makes up only about 0.04 percent of our atmosphere — the process necessitates large expenditures for energy and big equipment. What he has likewise observed, in analyzing similar industries that separate gases, suggests that translating spreadsheet projections for capturing CO2 into real-world applications will reveal hidden costs. “I think there has been a lot of hype about this, and it’s not going to revolutionize anything,” he told me, adding that he thinks other negative-emissions technologies will prove cheaper. “At best it’s going to be a bit player.”

Last year, when David Keith and his associates at Carbon Engineering published figures projecting that their carbon-capture technology could bring costs as low as $94 a metric ton, Herzog was not convinced. Keith nevertheless made the case to me that two new investors in Carbon Engineering — Chevron Technology Ventures and a subsidiary of Occidental Petroleum — scrutinized his company’s numbers to an exhaustive degree and agreed the economics of the venture were solid enough to merit putting up substantial amounts in a $60 million investment round. Both Climeworks founders told me they agreed with Keith’s cost estimates, and saw a similar downward curve for their own technology.

Climeworks’s current goal is to remove 1 percent of the world’s annual CO2 emissions by the mid 2020s. Yet meeting such a benchmark, if it’s even possible, would require bringing the cost of direct air capture down by nearly an order of magnitude while maintaining and expanding their roster of clients substantially. At the moment, Wurzbacher and Gebald have planned for several generations of Climeworks machines, with each new model promising declining prices. “Basically, we have a road map — $600, down to $400, down to $300 and $200 a ton,” Wurzbacher said. “This is over the next five years. Down to $200 we know quite well what we’re doing.” And beyond $200, Wurzbacher suggested, things get murkier. To move below that price would depend on “new developments” in technology or manufacturing.

Both founders told me they expect to reap enormous cost reductions from expanding production — activities that involve buying materials more cheaply in bulk and assembling units on automated factory lines instead of building them by hand, as is the case now. Design advances could wring out other costs. “Maintenance is very expensive,” Wurzbacher said. "Right now, if we exchange the filters in the collectors, we have to rent a crane, and that's a lot of man-hours. In the next-generation units, we have improved that a lot, so relatively small design changes could cut the costs of maintenance by a factor of three." Climeworks also intends to derive savings from improvements to crucial materials, like the sorbent that catches the CO2. At the moment, the company's technology requires that the temperature inside the units be raised periodically to about 100 degrees Celsius to release CO2 from the sorbent so it can be drawn off and stored. If the process can be done at a lower temperature, the units will use less energy, and the life of the materials should be extended, further driving down costs.

The company's ambitions for mass production may still seem extreme. To actually capture 1 percent of the world's carbon emissions by 2025 would, by Gebald's calculations, require that Climeworks build 250,000 carbon-capture plants like the ones on the roof at Hinwil. That adds up to about 4.5 million carbon collectors. For a company that has only built 100 collectors (and has 14 small plants around Europe), it's a staggering number. The Climeworks founders therefore try to think of their product as the automotive industry might — a piece of mass-produced technology and metal, not the carbon they hope to sequester. "What we're doing is gas separation," Wurzbacher said, "and that's traditionally a process-industry business, like oil and gas. But we don't really see ourselves there."

The founders note that Toyota makes more than 10 million cars annually. "Every CO2 collector has about the same weight and dimensions of a car — roughly two tons, and roughly 2 meters by 2 meters by 2 meters," Gebald said. "And all the methods used to produce the CO2 collectors could be well automated. So we have the automotive industry as a model for how to produce things in large quantities for low cost." The two men have already sought advice from Audi. They are also aware that the automotive industry perfected its methods over the course of 100 years. Climeworks, if it plans to have even a modest impact, doesn't have nearly as much time.

In 1954, the economist Paul Samuelson put forward a theory that made a distinction between "private-consumption goods" — bread, cars, houses and the like — and commodities that existed apart from the usual laws of supply and demand. Modern global markets are obviously quite successful at pricing private goods we need and want. But the other type of commodity Samuelson was describing is something now known as a "public good," which benefits everyone but is not bought, sold or consumed the same way. Definitions of a public good can vary, but the oft-used examples are lighthouses, national defenses and clean air.

Direct air capture can no doubt create private goods, like soft-drink carbonation or fuels. What makes its value so difficult to estimate is that in burying CO_2 for a better atmosphere — and, almost certainly, a better future — its purveyors would also create a public good. "The challenge with just collecting and burying CO_2 is that there isn't a market yet," Julio Friedmann, a former United States Energy Department official who now works at Columbia University, told me. "What it's really about is offering an environmental service for a fee." And what that means, in short, is that direct air capture's success would be limited to the size of the market for private goods — soda fizz, greenhouse gas — unless governments decided to intervene and help fund the equivalent of several million (or more) lighthouses.

An intervention could take a variety of forms. It could be large grants for research to find better sorbent materials, for instance, which would be similar to government investments that long ago helped nurture the solar- and wind-power industries. But help could also come by expanding regulations that already exist. A new and obscure United States tax provision, known as 45Q and signed last year by President Trump, offers a tax credit of up to $50 a ton for companies that bury CO_2 in geologic formations. The credit can benefit oil and gas firms that pump CO_2 underground during drilling work, as well as power plants that capture emissions directly from their smokestacks. Yet it could be used by Climeworks too, should it open plants in the United States — but only if it manages to remove and bury 100,000 tons of CO_2 per year.

Governments can make carbon more expensive too. The Climeworks founders told me they don't believe their company will succeed on what they call "climate impact" scales unless the world puts significant prices on emissions, in the form of a carbon tax or carbon fee. "Our goal is to make it possible to capture CO2 from the air for below $100 per ton," Wurzbacher says. "No one owns a crystal ball, but we think — and we're quite confident — that by something like 2030 we'll have a global average price on carbon in the range of $100 to $150 a ton." There is optimism in this thinking, he admitted; at the moment, only a few European countries have made progress in assessing a high price on carbon, and in the United States, carbon taxes have been repudiated recently at the polls, most recently in Washington State. Still, if such prices became a reality, they could benefit the carbon extraction market in a variety of ways. A company that sells a product or uses a process that creates high emissions — an airline, for instance, or a steel maker — could be required to pay carbon-removal companies $100 per metric ton or more to offset their CO2 output. Or a government might use carbon-tax proceeds to directly pay businesses to collect and bury CO2. In the absence of any meaningful government action, perhaps a crusading billionaire could put all the money in his estate toward capturing CO2 and stashing it in the earth.

If carbon came to be properly priced, a global ledger would need to be kept by regulators so that air-capture machines could suck in and bury an amount equivalent to the CO2 that emitters produce. Because CO2 emissions mix quickly into the atmosphere, location would be mostly irrelevant, except for the need to situate plants near clean energy sources and suitable areas for sequestering the gas underground. A direct-air-capture plant in Iceland, in other words, could take in the same quantity of emissions produced by a Boeing 787 in Australia and thus negate its environmental impact. What's more, there might not be limitations on the burial process. "It doesn't cost too much to pump CO2 underground," Stanford's Sally Benson says. Companies already sequester about 34 million metric tons of CO2 in the ground every year, at a number of sites around the world, usually to enhance the oil-drilling process. "The costs range from $2 to $15 per ton. So the bigger cost in all of this is the cost of carbon capture." Benson told me that various studies suggest that the earth's capacity for CO_2 sequestration could be in the range of 25 trillion metric tons; burying, say, five billion metric tons of CO_2 a year is therefore within the realm of possibility.

In an imaginary, zero-carbon future, the revenue prospects for air-capture companies would probably be enormous. "If we get to $100 to $150 a ton," Wurzbacher told me, "then the market is almost infinite." It would be so large, he said, that even if his company went through an exponential expansion, he doubted it could serve all the potential clients. At such low prices, companies could potentially fold carbon offsets into their pricing — or be compelled to do so — leading to an explosion in the market. "Christoph and me, we are always saying, we think that if this develops in a direction we think it does, we are not founding a company — we're really founding a new industry," Wurzbacher said. He points to the work in Iceland — a collaborative effort, funded partly by the European Union — as the first step toward that industry. At the moment, a single Climeworks collector on a Reykjavik geothermal field takes in air and collects CO_2; after the gas is flushed from the machine's filter, it is mixed with water, essentially forming hot seltzer. Then the liquid is injected into a basalt rock formation deep underground. Over the course of about two years, the CO_2 mineralizes, locking away the gas forever.

At Climeworks's offices in Zurich, I asked Valentin Gutknecht, who was at the time the company's business-development manager, if he could bury in Iceland my emissions from my plane flight from the United States to Zurich. He had a written agreement he could print out and give me, but it wouldn't be cheap, he warned. The price was running about $600 a metric ton, meaning my flight would cost about an extra $700. But I was hardly the first person to ask him. The weekend before, Gutknecht told me, he received 900 unsolicited inquiries by email. Many were from potential customers who wanted to know how soon Climeworks could bury their CO2 emissions, or how much a machine might cost them. I had the sense I was getting a glimpse of what's to come: A community of people — not large enough to make a difference, but nonetheless motivated — seemed ready to pay a premium to reverse their CO2 emissions.

Later, Wurzbacher told me he wants to offer a "one click" consumer service, perhaps in a year or two, which would expand what they're doing in Iceland to individual customers and businesses. A Climeworks app could be installed on my smartphone, he explained. It could then be activated by my handset's location services. "You fly over here to Europe," he explained, "and the app tells you that you have just burned 1.7 tons of CO2. Do you want to remove that? Well, Climeworks can remove it for you. Click here. We'll charge your credit card. And then you'll get a stone made from CO2 for every ton you sequester." He sat back and sighed. "That would be my dream," he said.

Paradoxical though it may seem, it's probable that synthetic fuels offer a more practical path to creating a viable business for direct air capture. The vast and constant market demand for fuel is why Carbon Engineering has staked its future on synthetics. The world currently burns about 100 million barrels of oil a day. David Keith told me he thinks that by 2050 the demand for transportation fuels will almost certainly be modified by the transition to electric vehicles. "So let's say you'd have to supply something like 50 million barrels a day in 2050 of fuels," he said. "That's still a monster market."

Steve Oldham, Carbon Engineering's chief executive, added that direct-air-capture synthetics have an advantage over traditional fossil fuels: They won't have to spend a dime on exploration. "If you were a brand-new company looking to make fuel, the cost of finding and then extracting fossil fuel is going to be really substantial," he says. "Whereas our plants, you can build it right in the middle of California, wherever you have air and water." He told me that the company's first large-scale facility should be up and running by 2022, and will turn out at least 500 barrels a day of fuel feedstock — the raw material sent to refineries.

Climeworks perceives a large market for fuels, too. In a town near Zurich called Rapperswil-Jona, the firm has installed a collector in a small plant, run by the local technical university, to produce methane. In a room about the size of a shipping container, the Climeworks machine takes in CO2 through an air duct and sends it through a maze of pipes to combine it with hydrogen, which is derived from water using solar power. When I visited, the plant was a few weeks away from being operational, but the methane coming out of the works could replace gasoline in the engine of just about any car, bus or truck outfitted to run on natural gas. At a larger plant in Italy, Climeworks recently joined a consortium of European countries to produce synthetic methane that will be used by a local trucking fleet. With different tweaks and refinements, the process could be adapted for diesel, gasoline, jet fuel — or it could be piped directly to local neighborhoods as fuel for home furnaces.

From an economic standpoint, synthetic fuels could allow producers to plug into a huge existing infrastructure — refineries, gas stations, cars, planes, trucks, homes, ships — and replace a product already in demand with something arguably better. But the new fuels are not necessarily cheaper. Carbon Engineering aspires to deliver its product at an ultimate retail price of about $1 per liter, or $3.75 per gallon. What would make the product competitive are regulations in California that now require fuel sellers to produce fuels of lower "carbon intensity." To date this has meant blending gas and diesel with biofuels like ethanol, but it could soon mean carbon-capture synthetics too.

In an expanding market, synthetic fuels could have curious effects. Since they're made from airborne CO2 and hydrogen and could be manufactured just about anywhere, they could rearrange the geopolitical order — tempering the power of a handful of countries that now control natural-gas and oil markets. The methane project in Rapperswil-Jona is especially suited for that country's needs, Markus Friedl, a thermodynamics professor overseeing the project, told me, because Switzerland imports almost all of its natural gas, and its ability to generate energy from renewable sources is limited during the colder months. Carbon-capture-derived fuels, if they become cheap enough, could be a form of energy storage — made in summer, with solar or wind power, and used in winter — that carries a lower cost (and longer life) than batteries.

From an environmental standpoint, air-capture fuels are not a utopian solution. Such fuels are carbon neutral, not carbon negative. They can't take CO2 from our industrial past and put it back into the earth. If all the cars, trucks and planes of the year 2050 run on renewable fuels instead of fossil fuels, their CO2 emissions would need to be removed from the air, recycled into the same product they originally burned through, and the cycle would need to repeat, ad infinitum, lest emissions increase. Even so, these fuels could present an enormous improvement. Transportation — currently the most significant source of emissions by sector in the United States — could cease to be a net emitter of CO2. Just as crucial, the technology of direct air capture could scale up to become better and cheaper.

A huge expansion would also involve huge complications. "You start to get into really big challenges when you get to these big, large scales," Glen Peters, a research director at the Cicero Center for International Climate Research in Oslo, told me. "If you can do one carbon-capture facility, where Carbon Engineering or Climeworks can build a big plant, great. You need to do that 5,000 times. And to capture a million tons of CO2 with direct air capture, you need a small power plant just to run that facility. So if you're going to build one direct-air-capture facility every day for the next 30 years to get to some of these scenarios, then in addition, we have to build a new mini power plant every day as well." It's also the case that you have to address two extraordinary problems at the same time, Peters added. "To reach 1.5 degrees, we need to halve emissions every decade," he said. That would mean persuading entire nations, like China and the United States, to switch from burning coal to using renewables at precisely the same time that we make immense investments in negative-emission technologies. And Peters pointed out that this would need to be done even as governments choose among competing priorities: health care, education and so on.

"The idea of bringing direct air capture up to 10 billion tons by the middle or later part of the century is such a herculean task it would require an industrial scale-up the likes of which the world has never seen," Princeton's Stephen Pacala told me. And yet Pacala wasn't pessimistic about making a start. He seemed to think it was necessary for the federal government to begin with significant research and investments in the technology — to see how far and fast it could move forward, so that it's ready as soon as possible. At Climeworks, Gebald and Wurzbacher spoke in similar terms, asserting that the conversations around climate challenges are moving beyond the choice between clean energy or carbon removal. Both will be necessary.

Gebald and Wurzbacher seem less assured about the future of global policy than on the mechanics of scaling up. Some of that, they made clear, was related to their outlook as engineers, to what they've gathered from observing companies like Audi and Apple. If the last century has proved anything, it's that society is not always intent on acting quickly, at least in the political realm, to clean up our environment. But we've proved very good at building technology in mass quantities and making products and devices better and cheaper — especially when there's money to be made. For now, Gebald and Wurzbacher seemed to regard the climate challenge in mathematical terms. How many gigatons needed to be removed? How much would it cost per ton? How many Climeworks machines were required? Even if the figures were enormous, even if they appeared impossible, to see the future their way was to redefine the problem, to move away from the narrative of loss, to forget the multiplying stories of dying reefs and threatened coastlines — and to begin to imagine other possibilities.

JON GERTNER writes frequently for the magazine about science and technology. He last wrote about Tesla's effort to build self-driving cars.

CHAPTER 6

Action and Inaction in the Political Sphere

Caution and denial have left most U.S. climate commitments unfulfilled. 2001 and 2017 saw presidents Bush and Trump, respectively, withdraw from international climate agreements. During his presidency, Obama attempted to encourage businesses to adapt with cap-and-trade policies, without success. Despite some progressive action on the part of state governments, the gains have been minimal. However, ambitious initiatives such as the Green New Deal suggest a path forward, while teen activists push to keep climate change in the public debate.

Bush Will Continue to Oppose Kyoto Pact on Global Warming

BY DAVID E. SANGER | JUNE 12, 2001

WASHINGTON, JUNE 11 — President Bush made clear today that he had no intention of reversing his opposition to a global warming accord supported by the European leaders he will meet with this week. And he strongly suggested that any new accord would have to bind developing nations, especially China and India, to the kind of commitments that would be made by the United States.

In an effort to mollify his European critics in the hours before he left for Spain tonight on his first trip to Europe as president, Mr. Bush acknowledged the severity of the global warming problem and said the United States would "lead the way by advancing the science on climate change." He described several new research initiatives that could mark a potentially significant focusing of American climate study.

But while suggesting a new approach to the issue of global warming, Mr. Bush remained firm in rejecting the 1997 Kyoto accord, noting that it set no standards for major emitters of greenhouse gases, like China and India, while creating mandates for the United States that could prove economically crippling. His aides further argued that the accord — aimed at reducing emissions of greenhouse gases below 1990 levels — was written to make it easier for Europe than for the United States to meet the goals.

Mr. Bush's outright rejection of the treaty two months ago led to an uproar in Europe. While unapologetic about their decision to back away from the accord, White House officials concede that they did a poor job of explaining their objections or their approach to the problem of reducing heat-trapping gases.

So today, Mr. Bush stepped into the Rose Garden with several of his cabinet members and publicly embraced a recent report from the National Academy of Sciences that concluded that temperatures are rising because of human activities. At the same time, he insisted that his rejection of the Kyoto protocol "should not be read by our friends and allies as any abdication of responsibility."

"We will act, learn and act again, adjusting our approaches as science advances and technology evolves," he said.

In essence, Mr. Bush was arguing that the market should be allowed to solve the problem, with the United States pushing along research "consistent with the long-term goal of stabilizing greenhouse gas concentrations in the atmosphere."

While advocating an attack on the problem of the buildup of greenhouse gases in the atmosphere, the president once again rejected the mandates in the Kyoto treaty that the United States and other developed nations cut their emission levels of those gases to well below

1990 levels, a move he said would be economically disastrous for the United States and the world. He offered no concrete alternatives to the Kyoto cutbacks, however, beyond research and the gradual application of new technology. And he reiterated his longstanding pledge that he would not agree to any accord that exempts the developing world. "The world's second largest emitter of greenhouse gases is China," Mr. Bush said, with Vice President Dick Cheney and Secretary of State Colin L. Powell at his side. "Yet China was entirely exempted from the requirements of the Kyoto protocol. India and Germany are among the top emitters. Yet India was also exempt from Kyoto."

Mr. Bush omitted any direct criticism of Europe, even though his aides have been saying, publicly and privately, that the members of the European Union have deliberately manipulated the debate — and unfairly caricatured Mr. Bush as an enemy of good environmental practice — to cover up their political problems coming into compliance with the Kyoto mandates.

Andrew Card, Mr. Bush's chief of staff, told reporters over lunch here today that the target of cutting greenhouse emissions to below 1990 levels was picked with "Machiavellian intent" because it enabled them to count in East Germany just before its economy was collapsing. One result is that Europe must now cut its emissions far less than the United States does, he argued.

Mr. Card argued that Mr. Bush had taken a courageous position that other nations would eventually come to appreciate. "The emperor of Kyoto was running around the stage for a long time naked," he said, "and it took President Bush to say, 'He doesn't have any clothes on.' "

Mr. Bush's statement today only seemed to fuel his disagreements with Europe, even as it was intended to tamp them down. "Everyone will be polite this week, I'm sure," said a senior European diplomat here, "but the standard everyone will be holding him to is how this stacks up against Kyoto. Where is the target? What is the U.S. timetable?"

Moreover, he has probably re-ignited the dispute with the developing world. China, for instance, has managed to reduce its emissions significantly in the last few years, and it argues that the United States has done comparatively little. Chinese officials have already said they view efforts to force stricter controls as part of a move to contain Chinese economic power.

Just as China and India have rejected limiting their economic potential by imposing strict environmental standards, Mr. Bush made clear today that he would not agree to any environmental limits that would slow the economy of either the United States or the world.

"We account for almost 20 percent of the man-made greenhouse emissions," he said. "We also account for about one-quarter of the world's economic output. We recognize the responsibility to reduce our emissions."

But he added that "we also recognize the other part of the story," saying the targets in the Kyoto treaty would "have a negative economic impact, with layoffs of workers and price increases for consumers."

Mr. Bush's statement was dismissed by a range of environmental groups as an effort to evade the issue by promising new scientific initiatives, but leaving unclear how much he was willing to spend, or how long the studies should take.

While Mr. Bush called for a "national climate change technology initiative" today, former members of the Clinton administration said it bore great resemblance to a $4.5 billion, five-year program they proposed four years ago. Congress never fully financed it, and Mr. Bush's recent budget did not support it.

"It's very weak tea," said David B. Sandalow, the former assistant secretary of state for oceans, environment and science and one of the negotiators of the Kyoto protocol in the last administration. Mr. Sandalow, now a senior fellow at the World Resources Institute, said, "If you were trying to develop a strategy to make sure China and India would not cooperate, you couldn't develop a better one than what Mr. Bush announced today."

What was striking about today's statement, though, was Mr. Bush's extensive discussion of the issue, and his commitment to do something about it — even as he swathed the specifics in a cloud of ambiguities.

He characterized global warming as a serious long-range problem but one whose dimensions were still too little understood.

He tacitly acknowledged that the United States' rejection of the Kyoto accord had estranged the United States from many nations with which it has good relations generally.

Accordingly, Mr. Bush said he would push for new efforts to study global warming and more coordination among research institutions. He called for more money to pay for research into ways to control greenhouse gases.

If some of the president's statements today about technology and America's own advances sounded familiar, it may be because it had echoes of his father's speech nine years ago this week at a major environmental conference in Rio de Janeiro that set the stage for the Kyoto negotiations.

"Let's face it, there has been some criticism of the United States," the first President Bush said at the time. "But I must tell you, we come to Rio proud of what we have accomplished and committed to extending the record on American leadership on the environment. In the United States, we have the world's tightest air quality standards on cars and factories, the most advanced laws for protecting lands and waters, and the most open processes for public participation."

He added, "Now for a simple truth: America's record on environmental protection is second to none."

In the years since, the United States has continued to support research and new technology and to push for limits on automobile exhaust and factory emissions.

By repeating his fidelity today to negotiating with other nations under the 1992 climate treaty signed by his father, Mr. Bush is essentially trying to reset the clock, arguing that Kyoto should be scrapped in favor of a new, market-based accord that did not impose such an onerous economic cost.

But it is far from clear that he can win any converts to that position. Eileen Claussen, president of the Pew Center on Global Climate Change, a nonpartisan group that works with many large corporations seeking to scale back their emissions, said today that she was "confused" about Mr. Bush's political goal.

"He is meeting with Europeans who are doing some very ambitious things to reduce emissions," she said. "Yet what we don't have from him is something that talks about how you go about reducing emissions."

Some of that, Bush administration officials said, is contained in his energy report, issued last month. Mr. Bush, for instance, called for the increased use of nuclear power, because it emits no greenhouse gases. In a sign of how far apart he and the Europeans are, Germany today reached an agreement with its utilities to phase out the use of nuclear power, in part because of the growing problem of disposing of nuclear waste.

Although this will be Mr. Bush's first trip to Europe since taking office, White House aides said today that the president, who critics have said has had little exposure to foreign countries, has made several previous trips to the region.

A White House spokesman, Gordon Johndroe, said Mr. Bush had been to the Britain at least three times, most recently in 1990, when he also visited Spain, Portugal and Morocco. He also said Mr. Bush had visited France, though no date was provided.

Growing Clamor About Inequities of Climate Crisis

BY STEVEN LEE MYERS AND NICHOLAS KULISH | NOV. 16, 2013

WARSAW — Following a devastating typhoon that killed thousands in the Philippines, a routine international climate change conference here turned into an emotional forum, with developing countries demanding compensation from the worst polluting countries for damage they say they are already suffering.

Calling the climate crisis "madness," the Philippines representative vowed to fast for the duration of the talks. Malia Talakai, a negotiator for the Alliance of Small Island States, a group that includes her tiny South Pacific homeland, Nauru, said that without urgent action to stem rising sea levels, "some of our members won't be around."

From the time a scientific consensus emerged that human activity was changing the climate, it has been understood that the nations that contributed least to the problem would be hurt the most. Now, even as the possible consequences of climate change have surged — from the typhoons that have raked the Philippines and India this year to the droughts in Africa, to rising sea levels that threaten to submerge entire island nations — no consensus has emerged over how to rectify what many call "climate injustice."

Growing demands to address the issue have become an emotionally charged flash point at negotiations here at the 19th conference of the United Nations Framework Convention on Climate Change, which continues this week.

At a news briefing here, Farah Kabir, the director in Bangladesh for the anti-poverty organization ActionAid International, described that country as a relatively small piece of land "with a population of 160 million, trying to cope with this extreme weather, trying to cope with the effect of emissions for which we are not responsible."

With expectations low for progress here on a treaty to replace the 1997 Kyoto Protocol, widely seen as having failed to make a dent in worldwide carbon emissions, some nations were losing patience with decades of endless climate talks, particularly those who see rising oceans as a threat to their existence.

"We are at these climate conferences essentially moving chess figures across the board without ever being able to bring these negotiations to a conclusion," Achim Steiner, executive director of the United Nations Environment Program, said in a telephone interview.

Although the divide between rich and poor nations has bedeviled international climate talks for two decades, the debate over how to address the disproportionate effects has steadily gained momentum. Poor nations here are pressing for a new effort that goes beyond reducing emissions and adapting to a changing climate.

While they have no legal means to seek compensation, they have demanded concrete efforts to address the "loss and damage" that the most vulnerable nations will almost certainly face — the result of fragile environments and structures, and limited resources to respond.

The sheer magnitude and complexity of the issue make such compensation unlikely. The notion of seeking justice for a global catastrophe that affects almost every country — with enormous implications for economic development — is not only immensely complicated but also politically daunting.

It assumes the culpability of the world's most developed nations, including the United States and those in Europe, and implies a moral responsibility to bear the costs, even as those same nations seek to draft a new treaty over the next two years that would for the first time compel reductions by rapidly emerging nations like China and India. As a group, developing countries will within a decade have accounted for more than half of all historical emissions, making them responsible for a large share of the continuing impact humanity will make, if not the impact already made.

Assigning liability for specific events — like Typhoon Haiyan, which struck the Philippines with winds of at least 140 miles an hour, making it one of the strongest storms on record — is nearly impossible. It can take scientists years just to determine whether global warming contributed to the severity of a particular weather event, if it can be determined at all.

Many negotiators here have pressed to create a new mechanism that effectively accepts the idea that the results of climate change are irreversible and that the countries that are hit hardest first must be compensated.

"We've reached a stage where we cannot adapt anymore," said Ronald Jumeau, the United Nations representative for the Seychelles, who is his country's chief negotiator here. He noted the devastating effects not only of extreme storm events, but also of creeping desertification, salinization and erosion that could result in financial losses and even territorial issues that the modern world has never had to face.

"This is new," he said. "This is like, 'The Martians are landing!' What do you do?"

John Kioli, the chairman of the Kenya Climate Change Working Group, a consortium of nongovernmental organizations, called climate change his country's "biggest enemy." Kenya, which straddles the Equator, faces some of the biggest challenges from rising temperatures. Arable land is disappearing and diseases like malaria are appearing in highland areas where they had never been seen before.

Developed countries, Mr. Kioli said, have a moral obligation to shoulder the cost, considering the amount of pollution they have emitted since the Industrial Revolution. "If developed countries are reasonable enough, they are able to understand that they have some responsibility," he said.

How to compensate those nations hardest hit by climate changes remains divisive, even among advocates for such action. Some have argued that wealthy countries need to create a huge pool of money to help poorer countries recover from seemingly inevitable losses of the tangible and intangible, like destroyed traditions.

Mr. Jumeau noted that Congress allocated $60 billion just to rebuild from one storm, Hurricane Sandy, compared with the $100 billion a year that advocates hope to see pledged to a Green Climate Fund by all nations. The fund, intended to help poorer countries reduce emissions and prepare for climate changes, has remained little more than an organizing principle since its creation in 2010, its fund-raising goals unmet.

Others have suggested a sort of insurance program.

The United States and other rich countries have made their opposition to large-scale compensation clear. Todd D. Stern, the State Department's envoy on climate issues, bluntly told a gathering at Chatham House in London last month that large-scale resources from the world's richest nations would not be forthcoming.

"The fiscal reality of the United States and other developed countries is not going to allow it," he said. "This is not just a matter of the recent financial crisis. It is structural, based on the huge obligations we face from aging populations and other pressing needs for infrastructure, education, health care and the like. We must and will strive to keep increasing our climate finance, but it is important that all of us see the world as it is."

Appeals to rectify the injustice of climate change, he added, will backfire. "Lectures about compensation, reparations and the like will produce nothing but antipathy among developed country policy makers and their publics," he said.

Juan Pablo Hoffmaister Patiño, a Bolivian who represents the alliance of developing nations known as the Group of 77 and China, said the issue was not so much about assigning culpability for the looming climate disaster as doing something to help those nations hardest hit.

"Trying to assign the blame is something that even scientifically could take us a very long time, and the challenges and problems are actually happening now," he said in an interview here. "And we need to begin addressing them now rather than identifying who is guilty and to what degree. We can't make this issue hostage to finding the responsible ones or not."

Meanwhile, global emissions continue to rise. A report this month by the United Nations Environment Program warned that immediate action must be taken to reduce emissions enough to limit the rise in average global temperatures to 2 degrees Celsius, or 3.6 degrees Fahrenheit, above preindustrial levels. That is the maximum warming that many scientists believe can occur without causing potentially catastrophic climate change.

The current global turbulence, consistent with what scientists expect to happen as the climate changes, is already taking a toll.

As the hundreds of diplomats and advocates assembled for talks here, Justus Lavi was waiting for rain in Kenya. The wheat, beans and potatoes he planted on his farm in Makueni County sprouted, but the rainy season brought only two days of showers, threatening to ruin his yield.

In northern Somalia, Nimcaan Farah Abdi's 10 acres of corn, tomatoes and other vegetables were ruined as violent storms swept the Horn of Africa. A typhoon last weekend in nearby Puntland killed more than 100 people, a disaster overshadowed by the far more destructive one in the Philippines.

"My farm has been washed away," Mr. Abdi said. It was the second year in a row of unusually heavy storms to have destroyed his livelihood, leaving him uncertain about how he will provide for his six children. "God knows," he added, "but I don't have anything to give now."

STEVEN LEE MYERS reported from Warsaw, and **NICHOLAS KULISH** from Nairobi, Kenya. **JUSTIN GILLIS** contributed reporting from New York, **DAVID JOLLY** from Paris, and **MOHAMMED IBRAHIM** from Mogadishu, Somalia.

Fossil Fuel Divestment Movement Harnesses the Power of Shame

BY DAVID GELLES | JUNE 13, 2015

NORWAY THIS MONTH became an unlikely leader in a growing social movement: persuading investors to sell their stock in fossil fuel companies.

In Norway's case, its $890 billion pension fund — the largest sovereign wealth fund in the world — will begin divesting itself of its stakes in coal companies. The move, approved by Parliament on June 5, offered a powerful endorsement of a tactic its backers say has the potential to reduce carbon consumption and in that way limit harmful greenhouse gas emissions.

The fossil fuel divestment movement, begun on the campus of Swarthmore College in Pennsylvania in 2011, has gathered force in only four years. AXA, the French insurance group, said it would sell $560 million in coal investments. The Rockefeller family said its enormous philanthropic arm would sell fossil fuel investments, starting with coal. And the endowments of several universities, including Stanford and Syracuse, have purged coal company stocks.

"There's been a tipping point in the last six months," Ben Caldecott who researches energy and climate change as director of the Stranded Assets Programme at Oxford University, said in an interview. Coal prices are being hammered. The Dow Jones coal index is down 86 percent since 2011. In a recent securities filing, Peabody Energy, one of the country's biggest coal producers, recently listed the divestment campaign as a risk factor that threatened its share price.

The logic of the campaign is that diminishing support from the markets will create financial hardship and ultimately lead fossil fuel producers to change. But there is an open secret: For all its focus on stock holdings, the true impact of divestment campaigns has nothing to do with a company's investor base, share price or creditworthiness.

GREG SAILOR FOR THE NEW YORK TIMES

The AEP Muskingum River power plant in Beverly, Ohio.

The most exhaustive study on the subject was conducted by researchers at Oxford, Mr. Caldecott among them. Their report, published in late 2013, examined previous divestment movements — like those against the government of South Africa in protest of apartheid, and against companies that sell tobacco, alcohol or pornography. It also looked closely at the emerging fossil fuel campaign, analyzing the targeted companies and their shareholders. The study concluded that even if every public pension fund and university endowment joined the movement and sold its fossil fuel stock, the effect would be negligible.

"The maximum possible capital that might be divested by university endowments and public pension funds from the fossil fuel companies represents a relatively small pool of funds," the study found.

That's largely because most energy company stock is held by big institutional investors like BlackRock and Fidelity, whose managers are unlikely to use their portfolios to advance moral or social agendas.

Moreover, few institutions vote with their dollars. During a three-decade divestment campaign against tobacco companies, only about 80 of organizations and funds ever sold stock to support the cause.

"Divestment in itself is neither here nor there," Atif Ansar, one of the study's authors and a professor at Saïd Business School at Oxford, said in an interview. "On its own, it's not going to generate any real impact."

That is, for all of the noise divestment campaigns create, they do little to affect the supply-and-demand economics that would undercut the business of mining, drilling for and refining fossil fuels. Even in the case of coal, the stocks of those companies are down not because of divestment, but because shale mining and cheap natural gas have reduced demand for coal.

But that does not mean divestment campaigns have no consequences. What they do best is good old-fashioned public shaming.

The Oxford researchers found that the negative publicity can create reputational headaches.

"It becomes much harder for stigmatized businesses to recruit good people, to influence policy and, occasionally, to raise capital," Mr. Caldecott said.

Divestment campaigns also give activists a focused — and easy to understand — object for their outrage.

"The goal is not to bankrupt the fossil fuel industry. We can't do that with divestment alone," said Bill McKibben, whose group, 350.org, is a leader in the divestment movement. "But we can help politically bankrupt them. We can impair their ability to dominate our political life."

Critics argue that damaging the reputations of coal and oil companies does nothing to reduce reliance on those fossil fuels.

"I'm very supportive of aggressive climate policies," said Robert Stavins, director of the Environmental Economics Program at the Harvard Kennedy School of Government. "But the message from the divestment movement is fundamentally misguided."

He contends that the problem is not investment in energy companies; it is an economy that remains dependent on fossil fuel production and consumption. While clean energy production is growing, Western economies would grind to a halt tomorrow without fossil fuels. And the divestment movement has focused on Western companies, while India and China have continued to mine and burn huge amounts of coal. Norway, now a leader in the movement, amassed its gargantuan sovereign fund by drilling for oil and gas in the North Sea.

"Divestment comes at the expense of meaningful action," said Frank Wolak, director of the Program on Energy and Sustainable Development at Stanford. "It will do nothing to reduce global greenhouse emissions. It will not prevent these companies from raising capital."

A more effective use of activists' energy, Mr. Wolak and Mr. Stavins said, would be to work on putting a price on carbon emissions through a carbon tax or a cap-and-trade system.

"What we need to do is focus on actions that will make a real difference," Mr. Stavins said, "as opposed to actions that may feel or look good, but have very little real world impact."

For divestment campaigners, moving markets is not really the point, despite their focus on stockholdings. Instead, they are more than happy to provoke companies like Exxon Mobil and Peabody Energy.

"If it polarizes the debate, it does so in a helpful way," Mr. McKibben said. "Left to their own devices, a sense of concern is inadequate to move the fossil fuel industry to action."

States and Cities Compensate for Mr. Trump's Climate Stupidity

EDITORIAL | BY THE NEW YORK TIMES | JUNE 7, 2017

PRESIDENT TRUMP'S DECISION to withdraw from the Paris compact on climate change was barely four days old when more than 1,200 governors, mayors and businesses promised to do whatever they could to help the United States meet the climate goals President Barack Obama had committed to in the agreement. In a letter, titled "We Are Still In," they declared that global warming imposes real and rising costs, while the clean energy economy to which the Paris agreement aspires presents enormous opportunities for American businesses and workers.

The statement was further evidence that Mr. Trump, as polls have shown, is out of touch with the American people. Yet this question remains: Can the United States meet its commitments without federal involvement? To many analysts, it's a hopeless task: Mr. Trump has not only removed America from a leadership role in the climate fight. He has also ordered his minions to kill or weaken beyond recognition every federal initiative on which Mr. Obama had based his pledge.

It would be unwise, however, to give in to pessimism.

Some context: The Paris agreement committed more than 190 nations to a collective effort to limit the rise in global warming to well below 2 degrees Celsius, or 3.6 degrees Fahrenheit, above pre-industrial temperatures. To that end, Mr. Obama promised to lower America's greenhouse gas emissions 26 percent to 28 percent below 2005 levels by 2025. That pledge was crucial to the overall goal; according to the think tank Climate Interactive, Mr. Obama's ambitious promise would account for one-fifth of the hoped-for global emissions reduction out to the year 2030.

Here's the good news: Thanks to market forces (chiefly the shift from dirty coal to cleaner natural gas); increased use of wind and solar power; more efficient vehicles, buildings and appliances; and aggressive state and local policies, emissions have already dropped

TOM HAUGOMAT

about 12 percent from 2005 levels, more than 40 percent of Mr. Obama's target. Further progress along these lines, without any new federal policies, would get us to a total emissions reduction of 15 percent to 19 percent by 2025, according to the Rhodium Group — way short of Mr. Obama's pledge. But if we could add back in the Obama initiatives — for instance, the mandatory shutdown of all old coal-fired power plants, which are rules Mr. Trump wants to kill — we'd get to 23 percent, which is much closer.

Mr. Trump, in short, has left a hole to fill. How to do it?

There are several pathways. First, state action: 29 states plus the District of Columbia have targets for how much of their electricity should come from renewable or alternative energy sources, and nine others have voluntary standards. Maryland and Michigan recently raised their targets. The nation's most populous state, California, is also its most ambitious. It is on track to get 33 percent of its electricity from renewable sources by 2020 and 50 percent by 2030. It also has a cap-and-trade system to put a price on emissions; Quebec is part of that system and Ontario will soon join. Other states ought to join that system, too.

Gov. Andrew Cuomo of New York and Gov. Jay Inslee of Washington have also set aggressive targets. Even red states like Iowa, Kansas and Texas have found that it makes economic sense to switch from coal to nonpolluting sources like wind, in part because costs of renewables have dropped sharply. Tucson Electric Power, an Arizona utility, recently announced that it would buy power from a solar farm for less than 3 cents a kilowatt-hour, which is less than half of what it paid in recent years. Experts say that price is comparable to the cost of power from plants fueled by natural gas. Utilities in North Carolina, Michigan and elsewhere plan to close coal-fired power plants.

Cities will also play a big role. New York and others are working to increase energy efficiency by updating building codes. They can also reduce emissions while improving public health by investing more in mass transit and electric vehicle charging stations. Businesses like Apple and Google say they intend to get all or nearly all of their energy from renewable sources. And who knows where electric vehicles will take us?

In 2008, the government projected that carbon emissions from power plants, industry, transportation and buildings would grow about 1 percent a year. Instead, they fell. Progress is possible, even with Mr. Trump standing in the way.

Fighting Climate Change? We're Not Even Landing a Punch

COLUMN | BY EDUARDO PORTER | JAN. 23, 2018

IN 1988, WHEN WORLD LEADERS convened their first global conference on climate change, in Toronto, the Earth's average temperature was a bit more than half a degree Celsius above the average of the last two decades of the 19th century, according to measurements by NASA.

Global emissions of greenhouse gases amounted to the equivalent of some 30 billion tons of carbon dioxide a year — excluding those from deforestation and land use. Worried about its accumulation, the gathered scientists and policymakers called on the world to cut CO2 emissions by a fifth.

That didn't happen, of course. By 1997, when climate diplomats from the world's leading nations gathered to negotiate a round of emissions cuts in Kyoto, Japan, emissions had risen to some 35 billion tons and the global surface temperature was roughly 0.7 of a degree Celsius above the average of the late 19th century.

It took almost two decades for the next breakthrough. When diplomats from virtually every country gathered in Paris just over two years ago to hash out another agreement to combat climate change, the world's surface temperature was already about 1.1 degrees Celsius above its average at the end of the 1800s. And greenhouse gas emissions totaled just under 50 billion tons.

This is not to belittle diplomacy. Maybe this is the best we can do. How can countries be persuaded to adopt expensive strategies to drop fossil fuels when the prospective impact of climate change remains uncertain and fixing the problem requires collective action? As mitigation by an individual country will benefit all, nations will be tempted to take a free ride on the efforts of others. And no country will be able to solve the problem on its own.

Still, the world's diplomatic meanderings — from the ineffectual call in Toronto for a reduction in emissions to the summit meeting in Paris, where each country was allowed simply to pledge whatever it could to the global effort — suggest that the diplomats, policymakers and environmentalists trying to slow climate change still cannot cope with its unforgiving math. They are, instead, trying to ignore it. And that will definitely not work.

The world is still warming. Both NASA and the National Oceanic and Atmospheric Administration reported last week that global temperatures last year receded slightly from the record-setting 2016, because there was no El Niño heating up the Pacific.

While the world frets over President Trump's decision to withdraw the United States from the Paris agreement, I would argue that the greatest impediment to slowing this relentless warming is an illusion of progress that is allowing every country to sidestep many of the hard choices that still must be made.

"We keep doing the same thing over and over again and expecting a different outcome," said Scott Barrett, an expert on international cooperation and coordination at Columbia University who was once a lead author of the Intergovernmental Panel on Climate Change.

Climate diplomats in Paris didn't merely reassert prior commitments to keep the world's temperature less than 2 degrees above that of the "preindustrial" era — a somewhat fuzzy term that could be taken to mean the second half of the 19th century. Hoping to appease island nations like the Maldives, which are likely to be swallowed by a rising ocean in a few decades, they set a new "aspirational" ceiling of 1.5 degrees.

To stick to a 2-degree limit, we would have to start reducing global emissions for real within about a decade at most — and then do more. Half a century from now, we would have to figure out how to suck vast amounts of carbon out of the air. Keeping the lid at 1.5 degrees would be much harder still.

Yet when experts tallied the offers made in Paris by all the countries in the collective effort, they concluded that greenhouse gas emissions in 2030 would exceed the level needed to remain under 2 degrees by 12 billion to 14 billion tons of CO_2.

Are there better approaches? The "climate club" proposed by the Yale University economist William Nordhaus has the advantage of including an enforcement device, which current arrangements lack: Countries in the club, committed to reducing carbon emissions, would impose a tariff on imports from nonmembers to encourage them to join.

Martin Weitzman of Harvard University supports the idea of a uniform worldwide tax on carbon emissions, which might be easier to agree on than a panoply of national emissions cuts. One clear advantage is that countries could use their tax revenues as they saw fit.

Mr. Barrett argues that the Paris agreement could be supplemented with narrower, simpler deals to curb emissions of particular gases — such as the 2016 agreement at a 170-nation meeting in Kigali, Rwanda, to reduce hydrofluorocarbon emissions — or in particular industries, like aviation or steel.

Maybe none of this would work. The climate club could blow up if nonmembers retaliated against import tariffs by imposing trade barriers of their own. Coordinating taxes around the world looks at least as difficult as addressing climate change. And Mr. Barrett's proposal might not deliver a breakthrough on the scale necessary to move the dial.

But what definitely won't suffice is a climate strategy built out of wishful thinking: the proposition that countries can be cajoled and prodded into increasing their ambition to cut emissions further, and that laggards can be named and shamed into falling into line.

Inveigled by three decades of supposed diplomatic progress — coupled with falling prices of wind turbines, solar panels and batteries — the activists, technologists and policymakers driving the strategy against climate change seem to have concluded that the job can be done without unpalatable choices. And the group is closing doors that it would do best to keep open.

There is no momentum for investing in carbon capture and storage, since it could be seen as condoning the continued use of fossil fuels. Nuclear energy, the only source of low-carbon power ever deployed at the needed scale, is also anathema. Geoengineering, like pumping aerosols into the atmosphere to reflect the sun's heat back into space, is another taboo.

But eventually, these options will most likely be on the table, as the consequences of climate change come more sharply into focus. The rosy belief that the world can reduce its carbon dependency over a few decades by relying exclusively on the power of shame, the wind and the sun will give way to a more realistic understanding of possibilities.

Some set of countries will decide to forget Paris and deploy a few jets to pump sulfur dioxide into the upper atmosphere to cool the world temporarily. There will be a race to develop techniques to harvest and store carbon from the atmosphere, and another to build nuclear generators at breakneck speed.

It will probably be too late to prevent the Maldives from ending up underwater. But better late than never.

EDUARDO PORTER writes the Economic Scene column for The New York Times. Economic Scene explores the world's most urgent economic encounters.

Reaching for a Zero-Emission Goal

BY PAUL RUBIO | SEPT. 21, 2018

COSTA RICA MAY BE SMALL and sparsely populated, but the Central American nation is a big player when it comes to environmental stewardship. In the late 1990s, Costa Rica emerged as a world leader in the ecotourism and sustainability movements — reversing decades of deforestation with successful initiatives to protect its land, seas and wildlife.

Now the country is tackling a far larger issue for its bicentennial in 2021: global climate change. Looking beyond national ecology, Costa Rica is implementing a series of new environmental policies to become Earth's first carbon-neutral nation. Below, we chart the progress achieved toward this lofty, zero-emission goal.

38 The age of Costa Rica's highly ambitious new president, Carlos Alvarado Quesada — a former journalist who is behind his country's carbon-neutral pledge.

300 The world record — set by Costa Rica in 2017 — for the most consecutive days of running electricity on renewable energy. Some 99 percent of the country's electricity originates from renewable resources (compared to 15 percent in the United States).

900 Members in Coopedota, which in 2011 became the first zero emission coffee company in Costa Rica — and the world. Since then, two more Costa Rican coffee companies have become zero-emission.

1994 When Costa Rica amended its constitution to incorporate the right to a healthy environment for all its residents.

31 Growth percentage in renewable energy investment in Costa Rica in 2016.

35 The passenger capacity of "Nyuti," a pioneering hydrogen-fueled passenger bus operating out of Liberia, Costa Rica (and that President

Carlos Alvarado Quesada took to his inauguration ceremony). This is the first clean energy bus in Central America.

2019 The year Costa Rica is set to gain its first carbon neutral airline, Green Airways. A major factor in gaining this neutrality is the airline's commitment to planting an endemic tree for every ticket purchased.

101 Companies in Costa Rica that are currently part of the public-private Alliance for Carbon Neutrality (Alianza para Carbono Neutralidad).

37,000 The number of electric cars the Costa Rican government hopes will be on its roads by 2022. To meet this goal, authorities plan to increase the electric car charging stations in the country to 61 from 20 by early 2019 and allow electric cars under $30,000 to be purchased tax-free.

25 Percent of Costa Rica's landmass officially protected as refuges, parks, reserves and other designated conservation areas. This includes 11 forest reserves, which serve as important carbon sinks.

309 Additional, underwater square miles granted protected status on World Oceans Day 2017 by the Costa Rican government. The new protected marine area of Cabo Blanco increases the nation's total percentage of protected seas to 15.7%.

A Green New Deal Is Technologically Possible. Its Political Prospects Are Another Question.

BY LISA FRIEDMAN AND TRIP GABRIEL | FEB. 21, 2019

WASHINGTON — President Trump derided the Green New Deal as a "high school term paper that got a low mark." Congressional Republicans mocked it as "zany." Even Nancy Pelosi, the Democratic House speaker, called the proposal a "green dream," and some of the party's 2020 candidates are starting to describe it as merely aspirational.

Yet, despite that disdain, the goals of the far-reaching plan to tackle climate change and economic inequality are within the realm of technological possibility, several energy experts and economists said in recent interviews.

Getting there will cost trillions of dollars, most agreed, and require expansive new taxes and federal programs. It certainly could not be accomplished within the 10-year time frame that supporters say is necessary, according to these experts.

The Green New Deal, in other words, is an exciting idea for many liberals and an enticing political target for conservatives. But, most of all, it is an extraordinarily complicated series of trade-offs that could be realized, experts say, with extensive sacrifices that people are only starting to understand.

Proposals for a Green New Deal — which would aim to slow climate change and catapult myriad industries into cutting-edge, low-carbon technologies — have been debated for more than a decade. But the subject was given new urgency last year by a high-profile United Nations report that said the Earth was on track to experience food shortages, fatal heat waves and mass die-offs of coral reefs by 2040, sooner than earlier projections. The report called for staggering changes to the global energy economy.

PETE MAROVICH FOR THE NEW YORK TIMES

Representative Alexandria Ocasio-Cortez of New York and Senator Edward J. Markey of Massachusetts, right, announcing the resolution on Feb. 7.

If the planet follows its current trajectory, the result by century's end would be "catastrophe," said John P. Holdren, the former science adviser to President Barack Obama. "The world would be almost unrecognizable compared to today's world."

"The evidence the climate is changing is becoming so overwhelming people are seeing it in their regions and in their lives," he added. "We are really to the point where we're seeing bodies in the street from severe flooding and severe wildfires."

The challenges in the Green New Deal for the economy, for Democratic presidential candidates running on it and for voters start with the fact that nearly 80 percent of America's energy now comes from relatively cheap and plentiful fossil fuels.

Replacing them with sources that do not emit greenhouse gasses will cost trillions of dollars; potentially increase energy costs for millions of families; and entail federal intervention in swaths of the economy, like transportation, where there is already a mixed record of government success. Republican critics gleefully noted last week that California's Democratic governor scaled back a state-owned bullet train linking San Francisco and Los Angeles because of costs.

Mitch McConnell, the Senate majority leader, has already said he will bring the plan to the floor, a move to force Democrats — particularly the six presidential candidates in the Senate who have endorsed the blueprint — to cast a vote that Republicans can use to brand them as socialists and extremists. Most 2020 hopefuls standing up for a Green New Deal have done little more than endorse it as a slogan and have rarely been pressed on its specifics.

But while the scope of the Green New Deal is enormous, experts believe that the economic trade-offs — saving trillions on potential catastrophe by spending trillions to prevent it — are worth serious consideration given the scale of the threat, and that a deep policy discussion would help voters and other Americans grapple with the environmental threats.

TECHNOLOGICAL CHALLENGES. AND POLITICAL ONES.

The Green New Deal, which is a congressional resolution without the force of legislation, calls for a "10-year national mobilization" to make the United States carbon-neutral across the economy. That means, as much carbon would have to be absorbed as is released into the atmosphere. Mr. Holdren, who is now a professor of environmental policy at Harvard University, said the Green New Deal's timeline of achieving that goal around 2030 is not feasible.

"As a technologist studying this problem for 50 years, I don't think we can do it," he said.

"There's hope we could do it by 2045 or 2050 if we get going now," he added.

Mr. Holdren said worldwide energy infrastructure — an investment of $25 to $30 trillion — turns over every three to four decades, and an aggressive transition to non-carbon energy begun today could achieve zero emissions by midcentury. That is the deadline urged by scientists from 40 countries in last year's report from the U.N.'s Intergovernmental Panel on Climate Change.

The more ambitious Green New Deal was introduced by Representative Alexandria Ocasio-Cortez of New York and Senator Edward J. Markey of Massachusetts. Its sweeping targets also include supplying 100 percent of the country's electricity from renewable and zero-emissions sources within a decade; digitizing the nation's power grid; upgrading every building to be more energy efficient; and overhauling factories and transportation, including cars, trucks and trains "as much as is technologically feasible" to remove greenhouse emissions.

The plan does not include a cost estimate, though it presumably would require massive new government spending and disrupt existing jobs and industries.

In addition to its climate goals, the plan includes far-ranging and politically problematic social promises, including guaranteed high-wage jobs, housing, paid vacation and health care.

Ethan Zindler, head of North American research at Bloomberg New Energy Finance, a clean-energy research group, said the power goal alone would be an enormous lift.

He noted that 37 percent of electricity in the United States comes from zero-carbon sources, including 20 percent of which is nuclear. If no new policies are enacted and all existing nuclear plants are kept online, the United States can rise to about 44 percent clean energy by 2030.

"We are quite optimistic that renewables will become the lowest-cost option" in the near future, Mr. Zindler said. But a 100 percent transformation to clean energy in a decade would necessitate not just shutting down coal, but also decommissioning natural gas plants.

"That would be extremely, extremely difficult to do verging on impossible without causing some real harm to the economy," he said.

Mark Z. Jacobson, a Stanford professor of civil and environmental engineering, was more optimistic. His research influenced a California law last year requiring the state to use 100 percent carbon-free

TOM BRENNER/THE NEW YORK TIMES

Senator Tom Cotton of Arkansas has said the deal would force Americans to have to "ride around on high-speed light rail, supposedly powered by unicorn tears."

energy by 2045. Mr. Jacobson said that 80 percent of the Green New Deal's target of net-zero greenhouse emissions across the economy could be achieved by 2030, and 100 percent between 2040 and 2050.

"You don't need any miracle technologies," he said.

He laid out a multistep plan: converting all energy to electricity and heat, and generating both solely with wind, solar and water resources; heating and cooling buildings with electric heat pumps; and powering factories with furnaces that use electricity. The only economic sector that can't be electrified with existing technology, he said, are long-distance airplanes and ships.

The Green New Deal "is technically and economically feasible," he said. "Socially and politically, it's a different question."

'TIME IS NOT OUR FRIEND'

Architects of the Green New Deal envision creating millions of high-wage jobs through its massive clean-infrastructure build-out. Labor unions have been cool to the plan, though, fearing that jobs in the renewable-energy sector won't be as high-paying or plentiful as those in oil and gas.

But Kelly Sims Gallagher, director of Tufts University's Center for International Environment and Resource Policy, said she believes the resolution's jobs goals are reachable.

There are now about 786,000 Americans working in the renewable energy industry, according to the most recent figures from International Renewable Energy Agency, compared to 3.8 million in China and 1.2 million in Europe.

Ms. Gallagher said the nation seems to have ceded some of these jobs to Europe and to China. "There's no reason why we couldn't get those back and build a stronger clean-energy industry in the United States," she said. Overhauling the transportation and buildings sectors within a decade are by far the biggest challenges the Green New Deal presents, she added. Both require major financial investments, regulations and — in the case of spurring electric-vehicle development and public transit — probably new taxes.

Ultimately, many experts said, it would not be possible to achieve Green New Deal goals without building into the economy a cost for emitting greenhouse gases — such as a carbon tax, which has long been a Republican, business-oriented approach.

Adele Morris, policy director of the climate and energy economics project at the Brookings Institution, said a greenhouse gas tax imposed on fewer than 3,000 taxpaying entities — corporations and municipal plants — would target 85 percent of United States emissions. And Daniel C. Esty a Yale environmental law professor and former commissioner of Connecticut's Department of Energy and Environmental Protection, said a $5-per-ton price on carbon that increases $5 each year over 20 years would put the United States in "full transformation mode" within a decade.

Republicans have mocked the Green New Deal as a "socialist wish list" untethered to economic realities. But the plan is popular with the Democratic base. A poll last week commissioned by environmental groups in early primary states — California, New Hampshire, South Carolina, Iowa and Nevada — found 74 percent of likely Democratic primary voters reacted favorably when the Green New Deal was described to them.

Although six Democratic presidential candidates are co-sponsors of the Green New Deal — Cory Booker, Kirsten Gillibrand, Kamala Harris, Amy Klobuchar, Bernie Sanders and Elizabeth Warren — it is unclear how familiar or supportive they are of its specifics. Most have offered general praise for its goals. Ms. Klobuchar, in an interview at a CNN town hall Monday, called the Green New Deal "so important right now for our country." But when pressed on whether the specific goals are achievable, she said, "I think that they are aspirations," and some compromises will be needed.

Mr. Sanders, who on Tuesday announced his candidacy for 2020, intends to release a plan for reaching the Green New Deal goals, said his spokesman, Josh Miller-Lewis.

In a recent interview Mr. Sanders said, "I'm prepared to be as bold as we can."

He said: "We are already spending many billions of dollars a year dealing with the impact of climate change," a figure certain to rise.

Mr. Booker, challenged by a Fox News reporter on Monday about the high costs to upgrade lighting alone, said, "This is the lie that's going on right now," while Ms. Warren, urging ambitious goals last week, said, "Republicans are stuck somewhere back in the 1950s."

The few congressional Republicans who want to address climate change have been wary of the Green New Deal. Representative Francis Rooney of Florida, the Republican co-chairman of the Climate Solutions Caucus, said he would likely vote against the resolution because of its lack of details and dismissal of "free enterprise and capitalism."

"I don't want to distract us from focusing on practical things we can actually accomplish like a carbon tax, like developing the infrastructure necessary to fight sea level rise," he said.

If the Green New Deal advances beyond a resolution to bill-writing, its policies have no chance of passing in the currently divided Congress, with a president who has mocked global warming as a hoax.

Thomas J. Pyle, president of the Institute for Energy Research, a pro-fossil-fuel group, pointed out that such policies could not even pass in Democratic strongholds like Washington State, where voters in November decisively rejected a ballot proposal for a carbon tax.

Mr. Pyle argued that Green New Deal boosters are not being realistic about the environmental consequences of constructing high-speed rail, manufacturing zero-emission vehicles or retrofitting buildings.

"How much steel is this going to involve? How much concrete? Think about the sheer amount of CO2 emitted into the atmosphere for retrofitting alone," he said. "It's almost as if they are suspending reality to get to their end goal."

But environmental activists said the details and hurdles are less important than the broad ambition of the plan, which proposes a national mobilization with the scale and urgency of the original New Deal.

"The science is clear: Time is not our friend here," said Carol Browner, a White House climate adviser to Mr. Obama and chief of the Environmental Protection Administration under President Bill Clinton. "So I have to say I'm as excited about this as I have been about anything in the environmental space in a long time."

LISA FRIEDMAN reported from Washington and **TRIP GABRIEL** from New York.

Students Across the World Are Protesting on Friday. Why?

BY AUSTIN RAMZY | MARCH 14, 2019

WHAT BEGAN AS one student's vigil calling for action on climate change has gone global, with school strikes planned in more than 100 countries on Friday. Here is a look at how the climate protests spread and how political leaders are responding.

A LONELY START TO AN INTERNATIONAL MOVEMENT

When a Swedish teenager, Greta Thunberg, sat before her country's Parliament in August, she was a solitary protester, armed with fliers that said she was refusing to attend school to protest adults' lack of concern for her future.

She sought to draw attention to the perilous state of the climate: Data from NASA has shown the past five years to be the warmest on record, and a report last year by the United Nations Intergovernmental Panel on Climate Change warned that without aggressive action, the world will face worsening wildfires, food shortages and other catastrophic effects as early as 2040.

Ms. Thunberg's protest quickly drew notice, with others joining in. After the Swedish elections in September, she made her strike a weekly event most Fridays. Then it began to spread, both online under the hashtags #climatestrike and #FridaysForFuture and on the streets.

In Australia, students from more than 200 schools skipped class on Nov. 30 to protest their country's climate policies and call for a ban on any new coal or gas projects.

A SHY TEENAGER BECOMES A LEADER

Ms. Thunberg, 16, is in some ways an unlikely figurehead for a worldwide movement.

An introvert prone to crippling depression who did not like to speak in class, she was powerfully shaken by lessons about pollution, species extinction and humans' influence on climate.

As her protest drew attention, she attended a United Nations climate conference in December in Poland, where she criticized negotiators. "You are not mature enough to tell it like it is," she said. "Even that burden you leave to us children."

In January, she traveled by train to attend the World Economic Forum in Davos, Switzerland, where she told a group of elites that many of them had made "unimaginable amounts of money" at the expense of the planet's future.

A Norwegian lawmaker said on Thursday that he and two colleagues had nominated Ms. Thunberg for the Nobel Peace Prize. She said she was "honored and very grateful."

Now the face of a global movement, Ms. Thunberg says her work has given her a welcome sense of purpose.

ELISABETH UBBE FOR THE NEW YORK TIMES

Greta Thunberg, who began a global youth movement calling for action on climate change, after a protest in Stockholm last month.

"I'm happier now," she told The New York Times last month. "I have meaning. I have something I have to do."

A CRITICAL RESPONSE FROM SOME LEADERS

Some school officials and politicians have criticized the protests, calling them a naïve misuse of class time.

Ms. Thunberg has responded sharply.

When Theresa May, Britain's prime minister, called the walkouts a waste of lesson time last month, Ms. Thunberg tweeted: "That may well be the case. But then again, political leaders have wasted 30 yrs of inaction. And that is slightly worse."

MORE THAN 1,600 EVENTS PLANNED WORLDWIDE

Last year, Ms. Thunberg called on the Swedish government to adopt policies in line with the Paris climate agreement, which sets a goal of limiting the global temperature rise from preindustrial levels to well below 2 degrees Celsius.

Students from other schools followed, and Friday's protests could be the largest yet. By early Thursday, more than 1,600 events were scheduled in at least 105 countries, according to organizers.

"Tomorrow we schoolstrike for our future," Ms. Thunberg tweeted. "And we will continue to do so for as long as it takes."

ILIANA MAGRA contributed reporting.

Glossary

anathema A topic, issue or option that is intensely opposed by a person or group.

attribution studies The scientific study of climate change's likely effect on a specific weather event.

cap-and-trade An approach to incentivize emissions reductions by placing carbon limits on firms, which are allowed to purchase credits for higher emissions from less polluting firms.

climate justice The goal of fighting social inequality along with fighting climate change, seen by activists and institutions as linked goals.

climate skepticism The mistrust or denial of consensus views of climate change.

deforestation The process of cutting down forests, either for logging or to open the land for other purposes.

direct air capture The technological process of removing carbon dioxide from the atmosphere.

divestment The process of selling off previous investments, often used as a form of protest against a business or country.

emission The release of a gas or radiation into the atmosphere.

Environmental Protection Agency A U.S. federal agency tasked with combating pollution and conserving natural resources.

feedback loop A term that refers to the end result of a process causing that process to continue.

fossil fuels High-carbon fuels that resulted from the buildup of decaying organic material over millions of years.

Green New Deal A proposed economic stimulus package that aims to address climate change and economic inequality.

greenhouse effect When particular atmospheric gases radiate light and heat, keeping the atmosphere warmer than it would be without their presence.

hybrid vehicle A vehicle that combines gas and electric power, lowering fuel requirements.

Keeling Curve A graph of the long-term concentration of carbon dioxide in the atmosphere, noted for its periodic seasonal variation and general upward trend.

Kyoto Protocol An international treaty obliging participant nations to commit to carbon reductions, signed by the United States in 1998 but never ratified due to resistance from President George W. Bush in 2001.

methane A relatively powerful greenhouse gas released by cattle industries, natural gas production and melting permafrost.

mitigation Efforts to reduce the severity of a problem, as opposed to solving it, now a component of climate action due to some of climate change's irreversible effects.

negative emission technologies All technologies that help remove carbon dioxide from the atmosphere, including direct air capture, enhanced weathering and other strategies.

ocean acidification The result of excess carbon dioxide being absorbed by oceans, making them more acidic and less hospitable to life.

Paris Agreement An international climate agreement, but not a formal treaty, that was signed by President Barack Obama and then withdrawn from by President Donald Trump.

turbine A device that uses rotation to convert kinetic energy into electricity, often using natural sources of movement like water or wind.

Media Literacy Terms

"Media literacy" refers to the ability to access, understand, critically assess and create media. The following terms are important components of media literacy, and they will help you critically engage with the articles in this title.

angle The aspect of a news story that a journalist focuses on and develops.

attribution The method by which a source is identified or by which facts and information are assigned to the person who provided them.

balance Principle of journalism that both perspectives of an argument should be presented in a fair way.

bias A disposition of prejudice in favor of a certain idea, person or perspective.

byline Name of the writer, usually placed between the headline and the story.

chronological order Method of writing a story presenting the details of the story in the order in which they occurred.

credibility The quality of being trustworthy and believable, said of a journalistic source.

editorial Article of opinion or interpretation.

feature story Article designed to entertain as well as to inform.

headline Type, usually 18 point or larger, used to introduce a story.

human interest story Type of story that focuses on individuals and how events or issues affect their life, generally offering a sense of relatability to the reader.

impartiality Principle of journalism that a story should not reflect a journalist's bias and should contain balance.

intention The motive or reason behind something, such as the publication of a news story.

interview story Type of story in which the facts are gathered primarily by interviewing another person or persons.

inverted pyramid Method of writing a story using facts in order of importance, beginning with a lead and then gradually adding paragraphs in order of relevance from most interesting to least interesting.

motive The reason behind something, such as the publication of a news story or a source's perspective on an issue.

news story An article or style of expository writing that reports news, generally in a straightforward fashion and without editorial comment.

op-ed An opinion piece that reflects a prominent individual's opinion on a topic of interest.

paraphrase The summary of an individual's words, with attribution, rather than a direct quotation of their exact words.

quotation The use of an individual's exact words indicated by the use of quotation marks and proper attribution.

reliability The quality of being dependable and accurate, said of a journalistic source.

rhetorical device Technique in writing intending to persuade the reader or communicate a message from a certain perspective.

tone A manner of expression in writing or speech.

Media Literacy Questions

1. "A Scientist, His Work and a Climate Reckoning" (on page 20) is an example of a human interest story. What kinds of details does the author use to appeal to readers?

2. What is the angle of "A Prophet of Doom Was Right About the Climate" (on page 37)? Compare it to "Global Warming Has Begun, Expert Tells Senate" (on page 10), written about the same individual thirty years earlier.

3. How does "Looking, Quickly, for the Fingerprints of Climate Change" (on page 47) attribute its sources? Identify the sources and if they are either directly quoted or paraphrased.

4. The headline of " 'Like a Terror Movie': How Climate Change Will Cause More Simultaneous Disasters" (on page 58) includes two parts. What is the purpose of each, and where does the first come from?

5. What is the intention of the article "India Is Caught in a Climate Change Quandary" (on page 80)? Compare the headline with the first sentence of the article as you think about your answer.

6. What is the tone of "The Amazon on the Brink" (on page 98)? What specific upcoming events would explain the reason for that tone?

7. The article "Hacked E-Mail Is New Fodder for Climate Dispute" (on page 107) reflects the conflicting perspectives of criticized scientists, climate skeptics and the mainline scientific position on climate change. What steps does the author take to treat the subjects impartially and truthfully?

8. "Energy Firms in Secretive Alliance With Attorneys General" (on page 111) uses political insiders and recently uncovered documents as sources. Why are such sources important for investigative reporting?

9. "An Energy Wish List for Congress" (on page 155) is an op-ed. Identify the authors' backgrounds from their bio. What details from their backgrounds inform their climate policy argument?

10. "The Tiny Swiss Company That Thinks It Can Help Stop Climate Change" (on page 159) is a feature story. How do the entertaining storytelling elements help convey the more technical details of the article?

11. "Growing Clamor About Inequities of Climate Crisis" (on page 183) is written in the inverted pyramid style. Compare the first and last paragraphs. How did the authors determine which information was most important?

12. The tone of articles on climate change is often strong: concerned, angry, dismissive or optimistic. What tone does "A Green New Deal Is Technologically Possible. Its Political Prospects Are Another Question." (on page 201) demonstrate?

Citations

All citations in this list are formatted according to the Modern Language Association's (MLA) style guide.

BOOK CITATION

THE NEW YORK TIMES EDITORIAL STAFF. *Adapting to Climate Change.* New York: New York Times Educational Publishing, 2020.

ONLINE ARTICLE CITATIONS

CHAN, SEWELL. "Paris Accord Considers Climate Change as a Factor in Mass Migration." *The New York Times*, 12 Dec. 2015, https://www.nytimes.com/2015/12/13/world/europe/paris-accord-considers-climate-change-as-a-factor-in-mass-migration.html.

DAVENPORT, CORAL. "E.P.A. Chief Doubts Consensus View of Climate Change." *The New York Times*, 9 Mar. 2017, https://www.nytimes.com/2017/03/09/us/politics/epa-scott-pruitt-global-warming.html.

EDDY, MELISSA. "Missing Its Own Goals, Germany Renews Effort to Cut Carbon Emissions." *The New York Times*, 3 Dec. 2014, https://www.nytimes.com/2014/12/04/world/europe/germany-carbon-emissions-environment.html.

FEARNSIDE, PHILIP, AND RICHARD SCHIFFMAN. "The Amazon on the Brink." *The New York Times*, 26 Sept. 2018, https://www.nytimes.com/2018/09/26/opinion/amazon-climate-change-deforestation.html.

FOUNTAIN, HENRY. "Calls for Shipping and Aviation to Do More to Cut Emissions." *The New York Times*, 16 Apr. 2016, https://www.nytimes.com/2016/04/17/science/calls-for-shipping-and-aviation-to-do-more-to-cut-emissions.html.

FOUNTAIN, HENRY. "Looking, Quickly, for the Fingerprints of Climate Change." *The New York Times*, 1 Aug. 2016, https://www.nytimes.com/2016/08/02/science/looking-quickly-for-the-fingerprints-of-climate-change.html.

FOUNTAIN, HENRY. "Scientists Link Hurricane Harvey's Record Rainfall to Climate Change." *The New York Times*, 13 Dec. 2017, https://www.nytimes.com/2017/12/13/climate/hurricane-harvey-climate-change.html.

FOUNTAIN, HENRY. "Why So Cold? Climate Change May Be Part of the Answer." *The New York Times*, 3 Jan. 2018, https://www.nytimes.com/2018/01/03/climate/cold-climate-change.html.

FRIEDMAN, LISA, AND TRIP GABRIEL. "A Green New Deal Is Technologically Possible. Its Political Prospects Are Another Question." *The New York Times*, 21 Feb. 2019, https://www.nytimes.com/2019/02/21/us/politics/green-new-deal.html.

GELLES, DAVID. "Fossil Fuel Divestment Movement Harnesses the Power of Shame." *The New York Times*, 13 June 2015, https://www.nytimes.com/2015/06/14/business/energy-environment/fossil-fuel-divestment-movement-harnesses-the-power-of-shame.html.

GERTNER, JON. "The Tiny Swiss Company That Thinks It Can Help Stop Climate Change." *The New York Times*, 12 Feb. 2019, https://www.nytimes.com/2019/02/12/magazine/climeworks-business-climate-change.html.

GILLIS, JUSTIN. "A Change in Temperature." *The New York Times*, 13 May 2013, https://www.nytimes.com/2013/05/14/science/what-will-a-doubling-of-carbon-dioxide-mean-for-climate.html.

GILLIS, JUSTIN. "Pace of Ocean Acidification Has No Parallel in 300 Million Years, Paper Says." *The New York Times*, 2 Mar. 2012, https://green.blogs.nytimes.com/2012/03/02/pace-of-ocean-acidification-has-no-parallel-in-300-million-years-paper-finds/.

GILLIS, JUSTIN. "A Prophet of Doom Was Right About the Climate." *The New York Times*, 23 June 2018, https://www.nytimes.com/2018/06/23/opinion/sunday/james-e-hansen-climate-global-warming.html.

GILLIS, JUSTIN. "A Scientist, His Work, and a Climate Reckoning." *The New York Times*, 21 Dec. 2010, https://www.nytimes.com/2010/12/22/science/earth/22carbon.html.

GILLIS, JUSTIN. "Why Should We Trust Climate Models?" *The New York Times*, 23 Sept. 2014, https://www.nytimes.com/news/un-general-assembly/2014/09/23/why-should-we-trust-climate-models/.

GILLIS, JUSTIN, AND HAL HARVEY. "An Energy Wish List for Congress." *The New York Times*, 5 Feb. 2019, https://www.nytimes.com/2019/02/05/opinion/clean-energy-climate-congress.html.

KANTER, JAMES. "A Shrinking Window for Burning Fossil Fuel." *The New*

York Times, 11 May 2009, https://green.blogs.nytimes.com/2009/05/11/a-shrinking-window-for-burning-fossil-fuel/.

LIPTON, ERIC. "Energy Firms in Secretive Alliance With Attorneys General." *The New York Times*, 6 Dec. 2014, https://www.nytimes.com/2014/12/07/us/politics/energy-firms-in-secretive-alliance-with-attorneys-general.html.

MYERS, STEVEN LEE, AND NICHOLAS KULISH. "Growing Clamor About Inequities of Climate Crisis." *The New York Times*, 16 Nov. 2013, https://www.nytimes.com/2013/11/17/world/growing-clamor-about-inequities-of-climate-crisis.html.

THE NEW YORK TIMES. "States and Cities Compensate for Trump's Climate Stupidity." *The New York Times*, 7 June 2017, https://www.nytimes.com/2017/06/07/opinion/climate-change-cities-states.html.

NIXON, RON. "Food Waste Is Becoming Serious Economic and Environmental Issue, Report Says." *The New York Times*, 25 Feb. 2015, https://www.nytimes.com/2015/02/26/us/food-waste-is-becoming-serious-economic-and-environmental-issue-report-says.html.

PINILLOS, N. ÁNGEL. "Knowledge, Ignorance and Climate Change." *The New York Times*, 26 Nov. 2018, https://www.nytimes.com/2018/11/26/opinion/skepticism-philosophy-climate-change.html.

PLUMER, BRAD. "Humans Are Speeding Extinction and Altering the Natural World at an 'Unprecedented' Pace." *The New York Times*, 6 May 2019, https://www.nytimes.com/2019/05/06/climate/biodiversity-extinction-united-nations.html.

PLUMER, BRAD. "Tropical Forests Suffered Near-Record Tree Losses in 2017." *The New York Times*, 27 June 2018, https://www.nytimes.com/2018/06/27/climate/tropical-trees-deforestation.html.

PLUMER, BRAD. "When Will Electric Cars Go Mainstream? It May Be Sooner Than You Think." *The New York Times*, 8 July 2017, https://www.nytimes.com/2017/07/08/climate/electric-cars-batteries.html.

PORTER, EDUARDO. "Fighting Climate Change? We're Not Even Landing a Punch." *The New York Times*, 23 Jan. 2018, https://www.nytimes.com/2018/01/23/business/economy/fighting-climate-change.html.

PORTER, EDUARDO. "India Is Caught in a Climate Change Quandary." *The New York Times*, 10 Nov. 2015, https://www.nytimes.com/2015/11/11/business/economy/india-is-caught-in-a-climate-change-quandary.html.

RAMZY, AUSTIN. "Students Across the World Are Protesting on Friday. Why?"

The New York Times, 14 Mar. 2019, https://www.nytimes.com/2019/03/14/world/europe/climate-action-strikes-youth.html.

REVKIN, ANDREW C. "Global Warming: Who Cares About a Few Degrees?" *The New York Times*, 1 Dec. 1997, https://www.nytimes.com/1997/12/01/world/global-warming-who-cares-about-a-few-degrees.html.

REVKIN, ANDREW C. "Hacked E-Mail Is New Fodder for Climate Dispute." *The New York Times*, 20 Nov. 2009, https://www.nytimes.com/2009/11/21/science/earth/21climate.html.

REVKIN, ANDREW C., AND MATTHEW L. WALD. "Material Shows Weakening of Climate Reports." *The New York Times*, 20 Mar. 2007, https://www.nytimes.com/2007/03/20/washington/20climate.html.

RUBIO, PAUL. "Reaching for a Zero-Emission Goal." *The New York Times*, 21 Sept. 2018, https://www.nytimes.com/2018/09/21/climate/costa-rica-zero-carbon-neutral.html.

SANGER, DAVID E. "Bush Will Continue to Oppose Kyoto Pact on Global Warming." *The New York Times*, 12 June 2001, https://www.nytimes.com/2001/06/12/world/bush-will-continue-to-oppose-kyoto-pact-on-global-warming.html.

SCHWARTZ, JOHN. " 'Like a Terror Movie': How Climate Change Will Cause More Simultaneous Disasters." *The New York Times*, 19 Nov. 2018, https://www.nytimes.com/2018/11/19/climate/climate-disasters.html.

SCHWARTZ, JOHN, AND HENRY FOUNTAIN. "Warming in Arctic Raises Fears of a 'Rapid Unraveling' in Region." *The New York Times*, 11 Dec. 2018, https://www.nytimes.com/2018/12/11/climate/arctic-warming.html.

SHABECOFF, PHILIP. "Global Warming Has Begun, Expert Tells Senate." *The New York Times*, 24 June 1988, https://www.nytimes.com/1988/06/24/us/global-warming-has-begun-expert-tells-senate.html.

SOBLE, JONATHAN. "Japan's Growth in Solar Power Falters as Utilities Balk." *The New York Times*, 3 Mar. 2015, https://www.nytimes.com/2015/03/04/business/international/japans-solar-power-growth-falters-as-utilities-balk.html.

TUGEND, ALINA. "On the Attack Against Climate Change." *The New York Times*, 21 Sept. 2018, https://www.nytimes.com/2018/09/21/climate/climate-change-groups.html.

WONG, EDWARD. " 'Irrational' Coal Plants May Hamper China's Climate Change Efforts." *The New York Times*, 7 Feb. 2017, https://www.nytimes.com/2017/02/07/world/asia/china-coal-gas-plants-climate-change.html.

Index

This book is current up until the time of printing. For the most up-to-date reporting, visit www.nytimes.com.